International Financial Markets
and Agricultural Trade

International Financial Markets and Agricultural Trade

EDITED BY
Thomas Grennes

Routledge
Taylor & Francis Group

LONDON AND NEW YORK

First published 1990 by Westview Press, Inc.

Published 2018 by Routledge
52 Vanderbilt Avenue, New York, NY 10017
2 Park Square, Milton Park, Abingdon, Oxon OX14 4RN

Routledge is an imprint of the Taylor & Francis Group, an informa business

Library of Congress Cataloging-in-Publication Data
International financial markets and agricultural trade / [edited by]
Thomas J. Grennes.
p. cm.
ISBN 0-8133-7811-7
1. Foreign exchange. 2. International
finance. 3. Produce trade.
I. Grennes, Thomas.
HG3851.I49 1990
382′.41—dc20   89-9062
                                                                    CIP

ISBN 13: 978-0-367-01424-7 (hbk)
ISBN 13: 978-0-367-16411-9 (pbk)

Contents

Preface

The papers contained in this volume originally were presented at a conference on International Finance and Agricultural Trade in San Antonio, Texas, in December 1988. The conference was sponsored by the International Agricultural Trade Research Consortium. Interest in the subject of the conference was stimulated by several recent events including (1) the twin budget and trade deficits of the United States, (2) large swings in the value of the dollar, and (3) the Uruguay Round of GATT negotiations that are focusing on agricultural trade.

I would like to acknowledge the assistance of several people who performed tasks related to the organization of the conference, including Nancy Schwartz of the Office of Management and Budget, David Orden of Virginia Polytechnic and State University, Karl Meilke of the University of Guelph, and Laura Bipes of the University of Minnesota. David Blandford of Cornell University presided as chairman of the Trade Consortium.

Publication of the book was supported by the Center for Economics and Business at North Carolina State University. Typing was ably done by Rita Edmond and editorial assistance was provided by Ann Phillips, both of the Department of Economics and Business at North Carolina State.

Thomas Grennes

1

The Link Between Financial Markets and World Agricultural Trade

Thomas Grennes

1. AGRICULTURAL TRADE

The agricultural sector in the United States and most other countries is strongly influenced by world economic conditions. Natural disturbances such as floods and drought are transmitted across national borders by trade. Changes in a large country's agricultural trade policy have a foreign impact as well as a domestic effect. In general, any forces that influence excess demands or supplies provide a link between national agricultural markets. Financial markets are an important part of the link between national product markets, particularly in the presence of floating exchange rates. Interest rates and exchange rates are the channels through which monetary and fiscal policy influence agricultural markets.

The average level of protection is higher for agricultural products than for manufactured goods as a result of exempting agriculture from the series of multilateral trade negotiations sponsored by the General Agreement on Tariffs and Trade (GATT). However, liberalization of agricultural trade is one of the main topics of the current Uruguay Round of GATT negotiations. Most high-income countries protect agriculture using a variety of instruments including tariffs, quotas, and export subsidies. In most cases the border measures are in place to prevent trade from interfering with domestic price support programs that transfer income to farmers. The United States has employed price support policies since the 1930s, and the result has been agricultural protectionism. In 1987 the Reagan administration made the remarkable proposal to the GATT to eliminate all trade-distorting subsidies to agriculture by the year 2000. The proposal has been resisted strongly by the European Community and Japan, and it remains to be seen whether a compromise will emerge. A factor in favor of liberalization is the high economic and budgetary cost of continued protection. In addition, competitive subsidization by the United States and the European Community has resulted in conditions that approach economic warfare.

Since exchange rates alter relative prices, they are related to protection. Currency devaluation is equivalent to a tariff on all imports and a subsidy on all exports. Thus, for a given level of tariffs, a country can change the level of agricultural protection by maintaining an overvalued or undervalued currency. For example, empirical studies of agricultural protection in low-income countries have found that a major source of negative protection is the use of overvalued currencies (World Bank 1986). Also, during the period from 1980 to 1985 when the dollar was rising in value, there was a large decrease in net agricultural exports of the United States.

Exchange rates and protection are inseparable in a world of floating exchange rates. Because of commodity arbitrage, domestic prices will change when exchange rates change even if foreign prices are constant. If a country commits itself to maintain a certain level of protection using any of the common measures (nominal protection, the tariff equivalent of quotas, or producer subsidy equivalent), the domestic price of the protected product must change in response to an exchange rate change.[1]

In the GATT negotiations there has been some discussion about what is the appropriate exchange rate and what is the appropriate foreign price for each product and country. The issue is important for agricultural policy because the volatility of exchange rates will influence the volatility of domestic prices. In the past the use of import quotas and variable levies has insulated domestic prices from the effect of exchange rate changes.

2. DEVELOPMENTS IN INTERNATIONAL FINANCIAL MARKETS

Floating exchange rates have prevailed in most countries since 1974, but the international monetary system continues to evolve. The optimum degree of coordination of monetary and fiscal policy remains an open issue. Other important developments are deregulation of financial markets in the United States and the opening of the Japanese financial market. Although floating rates have become the norm for national currencies, the European Monetary System has become a currency bloc with fixed rates among members whose currencies float jointly against those of nonmembers. The Delors Plan has called for closer monetary cooperation that would lead to a single Community central bank.

Dissatisfaction with floating rates has increased. Common complaints are excessive volatility of rates, misalignment of rates, and insufficient independence for the makers of monetary and fiscal policy. In an attempt to provide greater stability, central bankers of the

Group of Seven countries (United States, Japan, West Germany, France, Italy, Canada, and United Kingdom) apparently have agreed on a target range for exchange rates among major currencies. Although the existence of the Plaza Accord of 1985 and the Louvre Accord of 1987 has been widely reported in the press, the public never has been informed about what the target exchange rate range is, how it was established, or what factors might cause it to change. Presumably it is related to some notion of a confidence interval around an equilibrium exchange rate that can only be determined by some economic model. Unfortunately there is no consensus about what constitutes the appropriate model of exchange rate determination. However, in all prominent models, the equilibrium rate depends on monetary and fiscal policy in all countries. Hence, a range compatible with initial policies may not be compatible with subsequent policies.

For whatever reason, both nominal and real exchange rates have been more volatile since 1974 than during the earlier fixed-rate period. This increase in volatility may be an inherent feature of the floating exchange rate regime itself or it may be a result of greater volatility of the underlying variables such as monetary and fiscal policy. Many economic variables have been more volatile since 1974, including prices of primary products and prices of financial assets. During the floating rate period, exchange rates have not been more volatile than other asset prices or primary product prices. Also there has been no apparent increase in volatility since 1980 (see Pearce, chapter 7).

One possible effect of an increase in the volatility of exchange rates is a reduction in the volume of trade. In addition, overshooting of exchange rates or agricultural prices may act like a tax or a subsidy to the agricultural sector. This would make agriculture vulnerable to changes in monetary policy.

3. IMPORTANCE FOR TRADE OF REAL EXCHANGE RATES

To show the effect of exchange rates on trade, the inflation component must be removed from nominal exchange rates. Otherwise one would conclude mistakenly that currency depreciation caused by inflation would increase exports and increase imports. Countries like Argentina and Brazil do not improve their competitive situation when they experience a large nominal currency depreciation accompanied by a similar inflation rate. The demand for a country's exports depends on their prices in currency of the importing country, which depends on both the nominal exchange rate and the price in currency of the exporting country.

Several methods have been used for converting nominal rates into real exchange rates. The simplest and oldest method is to invoke Purchasing Power Parity (PPP). The real rate is then defined to be the nominal rate adjusted for the difference between domestic and foreign inflation rates. However, lack of empirical support for the accuracy of PPP in the short run has led people to seek an alternative measure.

Balassa (1964), Kravis and Lipsey (1983), and others have pointed out the problem created for PPP by the existence of a class of nontraded goods. This problem and others have led some people to use as a measure of the real rate the ratio of prices of traded goods to those of nontraded goods. Theoretical support for the idea comes from the Swan-Salter model of the dependent economy first developed in the 1950s. This measure of real rates and its relationship with the PPP measure is discussed in the chapters by L. Paul O'Mara and Sebastian Edwards.

Although bilateral exchange rates are useful for certain purposes, it is often convenient to construct an effective exchange rate index, which is an average rate between a particular country and the rest of the world. A weighted average must be constructed, but the weights can be based on a country's bilateral trade or they can be based on global trade weights. Edwards' paper finds the distinction between bilateral rates and multilateral rates to have some empirical importance. Other choices to be made are those between wholesale and consumer prices and the use of geometrical or arithmetic means.

Another issue involving weights is whether to include all traded goods or only those goods whose trade one is interested in analyzing. For example, the Economic Research Service of the U.S. Department of Agriculture publishes trade-weighted dollar indexes based on total agricultural trade and on trade for individual products such as wheat, corn, soybeans, and cotton. The Economic Research Service recently revised its dollar index published in the monthly Agricultural Outlook by including exchange rates for both countries importing from the United States and competing exporters (Stallings 1988).[2] An earlier index omitted competitor countries and gave the impression that changes in exchange rates between the United States, Canada, France, Australia, Argentina, and other exporters do not affect U.S. agricultural exports.

An observed exchange rate may be misaligned relative to its equilibrium value. Since an equilibrium value must be derived from an underlying economic model, there are many possible measures of misalignment. A direct application of PPP provides one measure (McKinnon 1988), and productivity-augmented versions of PPP also have been used (DeGrauwe and Verfaille 1988). Williamson (1985) has constructed a measure based on the notion of a "long-run sustainable current account balance." Edwards discusses these issues and presents his own model. He also considers the effect of misalignment on exports and economic growth.

4. PURCHASING POWER PARITY AND THE LAW OF ONE PRICE

PPP appears in both relative and absolute forms. Studies that disaggregate to narrow product categories tend to use the absolute version, often called the Law of One Price. Studies based on large product aggregates tend to use the relative version. Chapter two by Lawrence Officer considers these issues and the relationship between the aggregate and disaggregate studies.

A second issue involving PPP is its relative accuracy in the short run and the long run. Large and persistent short-run deviations have been found in nearly all recent studies. In their study of the adjustment process Frankel and Meese (1987) have estimated that from 9 to 14 percent of deviations from PPP have been eliminated in an average year.

A more fundamental criticism of PPP is that an exchange rate that moves away from equilibrium may never return to its initial real value. An efficient markets interpretation of PPP is that the real exchange rate follows a random walk (Roll 1979). Nominal exchange rates would not be forecastable and real rates should show no tendency to revert to their mean values. An alternative arbitrage version implies that real exchange rates should be mean-reverting in the long run. PPP would provide an anchor for the real rate. Thus, a time series of exchange rates should exhibit negative serial correlation. Frankel and Meese found some mean reversion in a study of the U.S. dollar-U.K. pound rate from 1869 to 1984. However, they claim that the post-1974 period of floating doesn't provide enough data to perform a satisfactory test of the mean reversion hypothesis.

5. THE LAW OF ONE PRICE AT DISAGGREGATE LEVELS

Officer in his survey of empirical studies of the Law of One Price at the disaggregate level finds only weak support for the hypothesis. He uses data from the International Comparison Project to relate prices of various product groups. The data are based on retail prices, and the largest deviations are for alcoholic beverages and tobacco. Since it is likely that these deviations are attributable to differential taxes, Officer suggests that a proper test would compare wholesale prices at the border. In addition to taxes, retail prices include tariffs, quotas, domestic transport costs, and marketing margins, and therefore large differences in retail prices are consistent with perfect commodity arbitrage. The literature on agricultural trade dealing with incomplete price transmission deals with several factors (for example, variable levies) that may prevent foreign price changes from being fully transmitted to domestic producers and consumers. Possible deviations from the Law of One Price due to incomplete commodity arbitrage are only one factor.

Goodwin's paper considers the specification of traditional tests of the Law of One Price that use contemporaneous prices. In his expectations-augmented model, prices in one market are compared with prices expected to prevail in a second market at the time of delivery. The results suggest that the traditional formulation is mis-specified.

6. EFFICIENCY OF THE FOREIGN EXCHANGE MARKET

Since exchange rate changes alter domestic agricultural prices, it is important to know whether the foreign exchange market efficiently processes information. Do exchange rate changes reflect changes in market fundamentals or merely noise? Douglas Pearce surveys the literature on international finance that borrows from the literature on stock market efficiency. Some of the same techniques have been applied to commodity markets, as shown by Pearce and his discussant, John Kitchen.

Several aspects of market efficiency are considered. Are exchange rates excessively volatile? Pearce presents data showing that exchange rates have not been more volatile than other asset prices or primary commodity prices, and that volatility has not increased since 1980. Variance bounds tests introduced by Shiller (1981) also are discussed.

Does the exchange rate follow a random walk? It need not in an efficient market if the underlying fundamentals do not follow a random walk. Monthly data are consistent with a random walk for the exchange rate, but daily data are not. An alternative approach to weak form efficiency is to search for trading rules that produce extraordinary profits.

Another aspect of information efficiency is whether the forward exchange rate is an unbiased forecast of the future spot rate. One difficulty in testing this proposition is that an efficient foreign exchange market may produce a biased forecast if there is a time-varying risk premium. Survey data have been used in an attempt to separate bias in forecasts from risk premia.

Market efficiency also can be assessed by considering whether prices react to new information. In particular, the unexpected portion of money supply announcements has been shown to affect interest rates and the stock market. Conversely, changes in the money supply that are anticipated by market participants should have no effect. Pearce also considers the effect of money announcements on exchange rates.

If foreign exchange markets fail tests of efficiency, what are the likely explanations? Three possibilities are offered by Pearce: (1) Expectations are not formed rationally, (2) rational speculative bubbles develop, and (3) government policy regimes change.

7. FISCAL POLICY, EXCHANGE RATES, AND TRADE

The effect of the recent budgetary and trade deficits of the United States are discussed by Douglas Purvis. During the period 1980-85, the dollar increased in value and the trade deficit increased. Also, the agricultural trade surplus of the United States decreased. Budget deficits have been offered as a possible explanation. However, the dollar decreased in value from 1985 to 1988. The decrease in the trade deficit and the increase in the agricultural trade surplus have been weaker than expected. Since budget deficits have remained large, it is difficult to attribute the declining dollar to fiscal policy. One problem is that theoretical models do not predict a definite sign for the effect of fiscal policy on exchange rates and empirical studies also show mixed results.

Purvis surveys the theoretical literature on the effects of fiscal policy on exchange rates. He includes the Mundell-Fleming model that incorporates capital mobility and emphasizes the short-run effects of aggregate demand. Extensions of the model considered include incorporations of the supply side of the economy, explicit treatment of expectations formation, and the treatment of portfolio balance.

A decrease in the trade deficit of the United States requires some combination of (1) an increase in private saving, (2) a decrease in private investment, or (3) a decrease in the budget deficit. Purvis considers the alternatives of a hard-landing brought on by a recession-induced decline in private investment and a soft-landing brought about by greater fiscal discipline and greater private saving.

NOTES

1. The relationship among the rate of protection (t), the domestic currency price (P), the foreign currency prices (P*), and the exchange rate (π) can be expressed as:

$$t = \frac{P}{\pi P*} - 1$$

2. An annual series for the agricultural trade-weighted dollar from 1960-87 appears in Stallings (1988, p. 23). A quarterly series from 1970 and a monthly series from 1975 are available from the Economic Research Service.

REFERENCES

Balassa, Bela (1964). "The Purchasing Power Parity Doctrine: A Reappraisal." *Journal of Political Economy* 72 (December): 584-96.

DeGrauwe, Paul, and Guy Verfaille (1988). "Exchange Rate Variability, Misalignment, and the European Monetary System," in Richard Marston (ed.), *Misalignment of Exchange Rates: Effects on Trade and Industry*. Chicago: University of Chicago Press.

Frankel, Jeffrey, and Richard Meese (1987). "Are Exchange Rates Excessively Variable?" in Stanley Fischer (ed.), *NBER Macroeconomics Annual 1987*. Cambridge, MA: MIT Press.

Kravis, Irving, and Robert E. Lipsey (1983). *Toward an Explanation of National Price Levels*. Princeton Studies in International Finance No. 52. Princeton, NJ.

McKinnon, Ronald (1988). "Monetary and Exchange Rate Policies for International Financial Stability." *Journal of Economic Perspectives* 2 (Winter): 83-103.

Roll, Richard (1979). "Violations of Purchasing Power Parity and Their Implications for Efficient International Commodity Markets," in M. Sarnat and G. Szego (eds.), *International Finance and Trade*. Cambridge, MA: Ballinger.

Shiller, Robert J. (1981). "Do Stock Prices Move Too Much To Be Justified by Subsequent Changes in Dividends?" *American Economic Review* (June): 421-36.

Stallings, David (1988). "Index Weighted by U.S. Agricultural Trade." *Agricultural Outlook* (October): 21-23.

Williamson, John (1985). *The Exchange Rate System*. Washington: Institute for International Economics.

World Bank (1986). *World Development Report*. Washington, D.C.

2

The Law of One Price:
Two Levels of Aggregation

Lawrence H. Officer

1. INTRODUCTION

The traditional statement of the law of one price (LOP) is that there is a unique price of a commodity or commodity basket worldwide, irrespective of the country of output or of absorption, where the respective domestic-country prices of the commodity or commodity basket are expressed in a common currency via market exchange rates. A reformulation of the law is presented below (Section 3), but the thrust of the paper is the review and coalescence of two hitherto disjoint strands of literature on the LOP: studies at the aggregate [gross-domestic-product (GDP)] level and those at a disaggregate level.

Interestingly, in both strands, the focus is less on the LOP itself than on the theoretical explanation and empirical testing of <u>deviations</u> from the LOP. There, however, the similarity ends; and conjunction of the approaches never begins. The present study, in contrast, after exploring the reasons for divergences from the LOP at the disaggregate level, uses this analysis for econometric explanation of deviations from the LOP at the GDP level. Further, a uniform data set is used to investigate the LOP at both levels of aggregation.

In Section 2 concepts of the price level--crucial in formulating and testing the LOP--are defined, with special attention paid to the underlying index-number measure, neglected in both branches of the literature. In Section 3 the LOP is formulated in terms of these price-level concepts and classified using a threefold schemata. The LOP at the disaggregate level is treated in Section 4 and at the GDP level in Section 5. In both sections existing approaches to testing the LOP are reviewed and their weaknesses discerned, enabling development of new techniques of investigating the law while avoiding these limitations. The techniques are applied to a commonly based data set in Sections 4 and 5. Conclusions of the study are presented in Section 6, and a data appendix follows.

2. PRICE-LEVEL CONCEPTS

2.1. Meaning of Price Level

In macroeconomic theory, a "price level" is the inverse of the commodity value of money. For a single commodity, the price level is its absolute (money) price. For a basket of commodities, it is a weighted average of the absolute prices of the individual commodities in the basket. So the price level in this sense has dimension "number of units of domestic currency per physical unit of commodity." In official statistics, the price level of a basket of commodities is always expressed as an intertemporal index number, representing the price level in the current period relative to a base period, in which case the price level is dimensionless.

In sharp contrast, the term "price level" as used in the present study involves an intercountry comparison. It is an interspatial relative price level: the price level of the commodity or commodity basket in the domestic country compared to its price level in another country (or group of countries). Just as for the domestic macroeconomy, the price level in this sense can be defined alternatively with dimension (yielding "absolute purchasing power parity") or without dimension (providing either "relative purchasing power parity" or the "real price level").

2.2. Purchasing Power Parity

Purchasing power parity (PPP) defines the intercountry relative price level as a ratio of respective own-currency price levels. Depending on whether the price comparison is at a point in time or over time, absolute PPP is distinguished from relative PPP. In either case, an underlying index-number concept must be selected (except for the special situation of a single commodity).

2.2.1. Absolute Form

The absolute PPP for a given domestic country (say, country j), denoted as PPP_j^A is the ratio of the number of units of currency j to the number of units of numeraire currency, each with the same purchasing power over the given commodity basket. Thus, absolute PPP has dimension "number of units of domestic currency per unit of numeraire currency." Following the terminology of Kravis, Heston and Summers (1982, p. 72, n. 2), a "numeraire" country or currency is merely a reference point, having only a scalar effect on the PPPs of a set of countries, while a "base" country has a distinct and nonproportionate influence on the PPPs.

Consider three alternative index-number concepts for PPP: Laspeyres, Paasche, and Geary-Khamis. Let p_{ij} (q_{ij}) denote the own-currency price (physical quantity, production or absorption) of commodity i in country j, and b (u) the subscript designation of the base (numeraire) country. Then, with m countries and n commodities, PPP_j^A is defined as follows for the respective indices:[1]

$$\underline{\text{Laspeyres}}: \quad PPP_j^A = \sum_{i=1}^{n} p_{ij}q_{ib} \Big/ \sum_{i=1}^{n} p_{ib}q_{ib} \tag{1}$$

$$j = 1, \ldots, m$$

$$\underline{\text{Paasche}}: \quad PPP_j^A = \sum_{i=1}^{n} p_{ij}q_{ij} \Big/ \sum_{i=1}^{n} p_{ib}q_{ij} \tag{2}$$

$$j = 1, \ldots, m$$

As a special case of both indices, for a single commodity i, $PPP_j^A = p_{ij}/p_{ib}$.

For the Laspeyres and Paasche indices, the base and numeraire countries are identical (b = u). For the Geary-Khamis index, equation (2) applies, but p_{ib} is the "international price" of commodity i and is defined as follows:

$$\underline{\text{Geary-Khamis}}: \quad p_{ib} = \sum_{j=1}^{m} (p_{ij}/PPP_j) \cdot (q_{ij}/\sum_{j=1}^{m} q_{ij}) \tag{3}$$

$$i = 1, \ldots, n$$

Thus the international price of a given commodity is the average of the individual-country prices each normalized by the PPP of the pertinent country and weighted by the relative quantity (output or absorption) of that country.

The Laspeyres index involves valuing the base-country's commodity basket alternatively at j-country and base-country prices and taking the ratio. The Paasche index values country j's commodity basket at its own and at base-country prices and takes the ratio. Both indices--as well as the Fisher "ideal" index, their geometric mean--possess the property of "characteristicity" in the sense that the price comparison (say, of the n-commodity basket underlying PPP_j^A between any two countries, say, j and b) is dependent on price and quantity data solely in these two countries.

In contrast, the Geary-Khamis index exhibits extreme noncharacteristicity, with the prices and quantities of all m countries involved in any two-country comparison. While the Geary-Khamis PPP_j^A is a Paasche index with "international" prices playing the role of "base-country" prices, the international prices are not known data. Rather, the international prices are defined in the n equations (3) and are obtainable only by solving the (m + n) equations (2), (3) simultaneously for the m PPPs and the n international prices.

It can be shown that the simultaneous-equation system has one degree of freedom, enabling imposition of the equation

$$PPP_u^A = 1 \tag{4}$$

and running the j subscript in (2) over only the remaining (m - 1) countries.[2]

With $PPP_j^A \quad \overset{n}{\underset{i=1}{R}} \ p_{iu}q_{iu} = \overset{n}{\underset{i=1}{R}} \ p_{ib}q_{iu}$, the dimensionality of the Geary-Kahmis remains "number of units of domestic currency per unit of numeraire currency," consistent with the other indices.

Equation (4) holds only for the n-commodity basket. For any given s-commodity subset of the basket, $PPP_j^A \neq 1$ and rather must be computed from equation (2) setting j = u and (rearranging the s commodities so they are numbered 1,...,s) running i over 1,...,s:

$$PPP_u^A = \overset{s}{\underset{i=1}{R}} \ p_{iu}q_{iu} / \overset{s}{\underset{i=1}{R}} \ p_{ib}q_{iu} \tag{5}$$

Then PPP_j^A given by (5) serves as a scalar factor enabling expression of PPP_j^A as a country-j/country-u comparison, thus retaining the above dimensionality:

$$PPP_j^A = (\overset{s}{\underset{i=1}{R}} \ p_{ij}q_{ij} / \overset{s}{\underset{i=1}{R}} \ p_{ib}q_{ij}) / PPP_u^A \tag{6}$$

2.2.2. Relative Form

The relative PPP of country j in period t with respect to period 0, PPP_j^R is the ratio of country j's intertemporal price index (PI_j) to the numeraire country's price index (PI_u), where each index is computed in period t with respect to the same base period 0 and for a similar commodity basket:[3]

$$PPP_j^R = PI_j/PI_u \qquad (7)$$

With each price index naturally dimensionless, their ratio (PPP_j^R) is also dimensionless. Just as for absolute PPP, various index-number concepts can underlie relative PPP. Customarily, however, only the Laspeyres (base-period-weighted) and Paasche (current-period-weighted) indices are computed in official statistics. Typical Laspeyres indices are consumer and wholesale price indices, while the GDP deflator and its components are examples of Paasche indices.

2.3. Real Price Level

The real price level (RPL) differs from PPP in expressing price levels or price indices in a common, numeraire currency. Letting E_j denote country j's market (nominal) exchange rate expressed as the number of units of currency j per unit of numeraire currency, the absolute RPL for country j, RPL_j^A , is defined as:

$$PPP_j^A = PPP_j^A/E_j \qquad (8)$$

and is the ratio of the number of units of currency j that exchanges for one unit of numeraire currency via the intermediary of a common commodity basket to the number of units of currency j obtainable directly in the foreign-exchange market. Alternatively, it can be interpreted as the ratio of country j's price level (the numerator of PPP_j^A) expressed in numeraire currency (via multiplication by $1/E_j$) to the numeraire-country's price level (the denominator of PPP_j^A). If PPP is at the GDP level, that is, if the basket of commodities (the q_{ij} over i for a given j) constitutes the entirety of GDP, then PPP_j^A is known as the "national price level" (although "domestic price level" better fits a GDP than a GNP concept).[4]

Correspondingly, country j's relative RPL, RPL_j^R, in period t relative to period 0 is the ratio of country j's price index re-expressed in numeraire currency to the numeraire-country's price index. Letting E_j^t (E_j^0) denote country j's exchange rate in period t (0),

$$RPL_j^R = [(E_j^0/E_j^t)PI_j]/PI_u = PPP_j^R/(E_j^t/E_j^0) \tag{9}$$

3. CLASSIFICATIONS OF THE LAW OF ONE PRICE

The LOP is the theory that the RPL is unity or less strictly (allowing for random error and even systematic forces that can cause RPL to diverge from unity) has a norm value of unity. A three-way classification of the LOP may be made. First, the LOP may be in either absolute or relative form (strictly, $RPL_j^A = 1$ or $RPL_j^R = 1$); that is, the theory may be that common-currency price levels are equal across countries at a point in time or that there has been an equiproportionate change in common-currency price levels since a base period.

Second, the price concept and weights underlying PPP and thence the LOP may pertain to either a final-expenditure (absorption) or an industry-of-origin (output) breakdown of GDP. Data for the former include the GDP-at-market-prices deflator, its expenditure component indices, and consumer price indices; for the latter, the GDP-at-factor-cost deflator, its industry component indices, and wholesale price indices (all of which, by example, pertain to relative PPP).

Third, the level of aggregation of the LOP--determined by the n-commodity or s-commodity aggregate used to compute PPP (see section on Absolute Form)--can vary considerably. At the highest level of aggregation is GDP itself; at the lowest is a precisely defined product, where not only does n = s = 1 but also the commodity exhibits extreme specificity and homogeneity. There are, of course, various levels between these extremes.

4. THE LAW OF ONE PRICE AT A DISAGGREGATE LEVEL

4.1. Theory

4.1.1. Justification for Norm Value of Unity

Clearly, absolute LOP is the basic theory and relative LOP derivative. Therefore the theoretical discussion will be confined to absolute LOP. What conditions make strict LOP hold ($RPL_i^A = 1$), thereby justifying a norm value of unity for RPL under less stringent circumstances? First, the relevant markets must be purely competitive in the Chamberlinian sense.[5] In particular, with the absence of monopoly elements, a given commodity is homogeneous across countries or, though formally differentiated, is perfectly substitutable. Second, the markets must be perfectly competitive (again using Chamberlinian terminology), meaning freely available and complete information, zero transport and transfer costs, and instantaneous adjustment.[6] In particular, commodity arbitrage is costless and instantaneous. Third, commodity markets must exhibit perfect efficiency, so that no profitable arbitrage opportunity prevails.

4.1.2. Reasons for Violations of the LOP

What forces move the RPL away from unity? First, there are deviations from perfect competition: real transport costs, including freight, insurance, and packing [Kravis and Lipsey (1971, p. 47)]; imperfect and costly information [Kravis and Lipsey (1971, p. 47), Johnson, Grennes and Thursby (JGT) (1979, p. 121), Bigman (1983, p. 81)]; interest foregone adding to arbitrage costs, because movement of goods takes time [Crouhy-Veyrac, Crouhy and Melitz (CCM) (1982, p. 328), Roll (1979, p. 139), Shapiro (1983, p. 301)]; arbitrageur's insistence on a risk premium because commodity prices and exchange rates may change during the shipping interval and forward or futures coverage may be unavailable for the desired duration or unduly expensive [Roll (1979, pp. 137-39), Jain (1980, p. 90), CCM (1982, p. 328), Shapiro (1983, pp. 302-303)].

Second, there may be divergences from pure competition <u>outside</u> the arbitraged-commodity market. National governments exert monopoly power by imposing tariffs and other trade restrictions. Even a perceived <u>probability</u> of a future trade barrier can restrain commodity arbitrage [JGT (1979, pp. 120-21), Jain (1980, p. 90)]. Further, governments engage in ostensibly domestic use of monopoly power by imposing indirect taxes on specific commodities [Genberg

(1975, p. 15)] and engaging in price-stabilization schemes (such as for agricultural commodities), which also adversely affect the LOP. Monopoly power can also be brought to bear by private firms outside the commodity markets. For example, shipping conferences and airline cartels can increase the expense of transportation beyond the real cost.

Third, the purity of competition may be absent in commodity markets even with perfectly homogeneous products. Suppliers can collude in their pricing or marketing [Kravis and Lipsey (1971, pp. 55-56)] or they can interact in noncolluding but nevertheless impurely competitive ways [Dornbusch (1987, pp. 96-98)]. Also, monopolists and oligopolists can engage in price discrimination between the domestic and foreign market or between various foreign markets [Kravis and Lipsey (1978, p. 228)].

Fourth, product differentiation directly reduces the substitutability of commodities between countries, thereby making competition impure [Kravis and Lipsey (1971, pp. 42-43, 57-59; 1978, pp. 203-204), Norman (1978, pp. 435-36), Isard (1977, p. 942), JGT (1979, p. 120)]. Including as commodity characteristics not only product-specific features (physical and performance attributes) but also supplier-specific elements (service attributes such as instruction in use, repair, reliability of supply, speed of delivery, and credit terms) enhances the importance of product heterogeneity as a reason for violations of the LOP [Kravis and Lipsey (1971, pp. 35, 59-60; 1978, pp. 203, 228-29), Fromm (1974, p. 400), JGT (1979, pp. 120-21)]. Further, product differentiation interacts with divergences from perfect competition and/or other divergences from pure competition to compound the impact of forces generating deviations from the LOP. For example, because of (i) the costs inherent in changing list prices or in deviating from them [Magee (1979, p. 157), Daniel (1986, p. 317)], or (ii) information costs required to recompute profit-maximizing prices [Daniel (1986, pp. 314-17)], or (iii) the simple desire for price stability [Dunn (1970, 1973)], oligopolistic firms may maintain the levels of prices expressed in the respective local currencies of their markets even in the face of exchange-rate changes.[7]

Fifth, commodity markets may be inefficient in the sense that not all available information is fully incorporated into prices and as a result, profitable arbitrage opportunities can persist.[8]

Sixth, a host of measurement problems may lead to spurious deviations from the LOP. (1) Product categories may not be consistent across countries, so that prices of comparable commodities are not being compared [Norman (1978, pp. 435-36), Officer (1986, p. 162)]. (2) The timing of price recordings may differ between countries

[JGT (1979, p. 121), Protopapadakis and Stoll (1986, p. 349)]. (3) Weighting patterns of ostensibly comparable price indices may differ between countries [Isard (1974, pp. 379-81), Kravis and Lipsey (1978, p. 199), Roll (1979, p. 148), Solnik (1979, p. 87), Shapiro (1983, p. 299), Dornbusch (1987, p. 1076)]. (4) Quality changes in commodities may be subject to differential treatment in different countries' official statistics [Genberg (1975, pp. 18-19), CCM (1982, p. 329, n. 6), Protopapadakis and Stoll (1983, p. 1433)]. (5) Official statistics often reflect "contract prices" (negotiated in the past), whereas only pure "spot prices" (current quotations) can be attuned to current exchange rates [Magee (1978; 1979, p. 157)].

4.2. Forces Enhancing the LOP

Commodity arbitrage combined with market efficiency is the obvious mechanism limiting deviations from the LOP in the face of impure and imperfect competition, but other forces exist. High elasticities of substitution in production and/or consumption minimize the adverse impact of product heterogeneity on the LOP. Neglected in the literature is the role of similarity of economies. If countries are similar in technology, factor supplies, market structure, demand conditions, and government policies, then--independent of the impurity and imperfection of competition and independent of the cost and efficiency of commodity arbitrage--the countries' RPLs will be close to unity. As an extreme case, if the economies are clones of each other, then strict LOP holds with no international exchange at all!

4.3. Previous Empirical Investigations

Table 2.1 summarizes all the existing empirical tests of the LOP at a disaggregate level of which I am aware. The highest level of aggregation, GDP, is excluded, as it is covered in Section 5. The next highest level, a division of output into tradables and nontradables is also omitted because it is treated elsewhere [Officer (1986)]. This leaves a variety of levels of aggregation, the set of potential levels determined by the classification scheme adopted by the individual author. Some researchers utilize the Standard International Trade Classification [United Nations (1986b)] with its five-digit (five-level) disaggregation scheme; others   select the   Standard   Industrial

TABLE 2.1. Tests of the Disaggregate Law of One Price

Study[a]	Countries[b]	Commodities No.[c]	Type[d]	Level of Aggregation	Nature of Price Data[e]	Result[f]
I. DIRECT COMPUTATION OF RPL (ABSOLUTE LOP)						
KL (1971)[g]	JAP GER UK US*	6	man	high	XPI	negative
Jain (1980)	UK US	24	primary	lowest	CEP	negative
CCM (1982)	FRA US	2	primary	lowest	WP	positive
PS (1983)	UK US[h]	15	primary	lowest	CEP	mixed
II. DIRECT COMPUTATION OF RPL (RELATIVE LOP)						
BC (1977)[i]	CAN US	18	man	low	WPI	negative
KL (1978)	GER US	9	man	low	XPI	negative
Magee (1979)	JAP US	4	man	low	WPI	negative
Dornbusch (1987)	GER JAP US*	28[j]	man	low	XPI	negative
"	US	12	man	low high	XPI MPI WPI	negative
"	US	9	man	low	MPI WPI	negative

Table 2.1 (continued)

Study[a]	Countries[b]	Commodities No.[c]	Type[d]	Level of Aggregation	Nature of Price Data[e]	Result[f]
III. RPL CHANGE COMPARED TO EXCHANGE-RATE CHANGE (RELATIVE LOP)						
Isard (1974)	GER US	11	man	high	XPI WPI	negative
Rosenberg (1974)	EUR JAP US*	4	man	lowest	MP WP	positive
Norman (1975)	UK	34	man	low	UVM WPI	negative
Isard (1977)	GER US	15	man	low high	XPI WPI	negative
"	CAN GER JAP US*	5	man	low	UVX UVM	negative
KL (1978)	GER US	8	man	low high	XPI	negative
Dornbusch (1987)	US	2	man	high	XPI MPI	negative
IV. B1-VARIABLE REGRESSION ANALYSIS (RELATIVE LOP)						
Dunn (1970, 73)	CAN US	6	both	low	WPI	negative
Curtis (1971)	CAN US	5	man	low	WPI	negative

Table 2.1 (continued)

Study[a]	Countries[b]	Commodities No.[c]	Type[d]	Level of Aggregation	Nature of Price Data[e]	Result[f]
Genberg (1975)	UK US OTH	6	primary	lowest	CEP	positive
Jain (1980)	UK US	24	primary	lowest	CEP	negative
Ormerod (1980)	UK	1	man	highest	UVX	negative
CCM (1982)	FRA US		both	lowest	WP	mixed
PS (1983)	UK US[h]	15	primary	lowest	CEP	mixed
Goodwin (1988)	JAP NETH US*	9	primary	lowest	CEP	mixed
V. TRI-VARIABLE REGRESSION ANALYSIS (RELATIVE LOP)						
Curtis (1971)	CAN US	50	man	low	WPI	negative
BC (1971)[i]	CAN US	18	man	low	WPI	negative
KL (1977)	GER US	2	man	highest	XPI WPI	negative
Richardson (1978)	CAN US	22	man[k]	low	WPI	negative
Jain (1980)	UK US	24	primary	lowest	CEP	negative

Table 2.1 (continued)

22

Study[a]	Countries[b]	Commodities No.[c]	Type[d]	Level of Aggregation	Nature of Price Data[e]	Result[f]
PS (1986)	UK US[h]	15	primary	lowest	CEP	mixed
Smith (1988)	ARG AU CAN THI US*	1[m]	primary	lowest	CEP	mixed
Goodwin (1988)	UK US	11	primary	lowest	CEP	mixed
VI. INVESTIGATIONS OF DIFFERENTIAL PRICING (RELATIVE LOP)						
KL (1974)[l]	GER JAP UK US	6	man	high	XPI WPI	negative
Ripley (1974)	JAP GER US	12[m]	both	varies	XPI UVX WPI	mixed
KL (1977)	GER US	2	man	highest	XPI WPI	negative
KL (1978)	GER US	1	man	highest	XPI WPI	negative
"	US	26	man	low	XPI WPI	negative
"	GER	69	man	low	XPI WPI	negative
Dornbusch (1987)	US	9	man	low	XPI WPI	positive

Table 2.1 (continued)

[a] Legend: BC = Bordo and Choudhry; CCM = Crouhy-Veyrac, Crouhy, and Melitz; KL = Kravis and Lipsey; PS = Protopapadakis and Stoll.

[b] Legend: AU = Australia; EUR = UK and European Economic Community; OTH = one or two major producing countries; THI = Thailand. Otherwise, obvious country abbreviations. A * superscript indicates the numeraire country.

[c] Number of commodity-specific price levels that are compared.

[d] Legend: both = manufactured goods and primary products; man = manufactured goods; primary = primary products.

[e] Legend: CEP = commodity-exchange or major-producing-country absolute price; MP = import absolute price; MPI = import price index; UVM = unit value of imports; UVX = unit value of exports; WPI = wholesale price index or industry unit-value index or industry selling-price index or producer price index; WP = same as WPI except absolute price rather than index; XPI = export price index.

[f] Legend: positive (negative) = favorable (unfavorable) to the LOP.

[g] Summarized in KL (1977, pp. 156–57; 1978, p. 230).

[h] Replaced by Australia and Malaysia for one commodity each.

[i] As summarized in Genberg (1978, pp. 250–51). Some entries are educated guesses.

[j] Average for the two-country comparisons.

[k] Predominantly.

[l] See also KL (1971, ch. 8) and, for a summary, KL (1978, p. 237).

[m] Average for each country.

Classification which for the United States [Office of Management and Budget (1987)] also incorporates five levels. Others use the eight-digit breakdown of the U.S. Bureau of Labor Statistics (1987, p. 191), and still others employ alternative schemes or must merge classification systems of various countries or use data of a nonclassificatory nature. In the fifth column of Table 2.1, a homogeneous treatment is provided by rating the level of aggregation of each study according to a four-step ordering--lowest, low, high, highest--where "lowest" denotes the finest level of disaggregation.

Five sets of studies corresponding to techniques of testing the LOP are distinguished in the table. First (groups I and II in the table), the RPL may be computed directly; test statistics consist of the mean and measures of variation around it, with the LOP involving values of one and zero, respectively.[9] It would be better to define the variation around the norm value, unity, but no author does so, although Magee (1979) implicitly adopts this principle in his measures. Second (group III in the table), changes in the RPL may be compared with changes in the nominal exchange rate; the LOP predicts no relationship.

The third and fourth techniques involve regression analysis. The three components of the RPL--the domestic price measure, foreign price measure, and exchange rate--are divided alternatively into a two-component multiplicative variable versus the remaining variable or into the three components, and one variable is regressed on the other (IV, "bi-variable regression analysis") or on the remaining two variables (V, "tri-variable regression analysis"). The LOP is not rejected if the constant term is not significantly different from zero and the slope coefficient(s) are (i) significantly different from zero, (ii) not significantly different from unity, and (iii) (though not all investigators state this criterion) in fact close to unity. The fifth approach (VI in the table) investigates firms' differential international pricing by comparing countrys' export and domestic prices as tests of discriminatory pricing. Price ratios are computed and tested for unity or correlations are calculated.

The most salient feature of Table 2.1 is the overwhelming preponderance of results unfavorable to the LOP. Of forty tests, twenty-eight reject the LOP, eight have mixed findings, and only four unambiguously cannot reject the LOP. However, rather than dismiss the LOP out of hand on the basis of this evidence, limitations of these tests suggest an improved methodology for investigating the LOP.

4.4. An Improved Empirical Methodology

1. Of the forty tests summarized in Table 2.1, thirty-six (groups II-VI) pertain to relative and only four (group I) to absolute LOP. Relative LOP suffers from the need to select a base period, an arbitrary decision except via the criterion that absolute LOP is satisfied in the base period, which cannot be known without testing for absolute LOP. Otherwise, even if relative LOP is found to hold, it could be perpetuating a base-period violation of absolute LOP. Correspondingly, a rejection of relative LOP might involve absolute LOP closer to fulfillment in the current than in the base period. A correct empirical methodology must involve absolute LOP as the version to be tested.

2. The bi-variable and tri-variable regression analyses involve a causal stance in identifying dependent and independent variable(s), whereas the LOP is a simple statement of a norm value of the RPL, regardless of how its actual value is generated. Further, only a model incorporating differential expectations of future commodity prices and exchange rates can justify tri-variable as distinct from bi-variable analysis--and now an even stronger causal model is involved.[10] Appropriate methodology would not decompose the RPL variable.

3. Investigations of differential (discriminatory) pricing, in focussing solely on individual countrys' prices, study only a subsidiary aspect of the LOP. Comparable commodity baskets produced (irrespective of destination) or absorbed (irrespective of source) in different countries are the essence of the LOP. A related comment applies to comparing changes in the RPL and the exchange rate. Again the focus is on a special case--oligopolistic sticky prices in the face of exchange-rate movements--rather than on the LOP in its basic form. Direct computation of the RPL is left as the technique of choice.

4. However, this approach as employed to date computes the wrong statistics. The mean algebraic deviation from the LOP generates a test overly favorable to the LOP, as positive and negative deviations cancel each other. The appropriate statistic is rather the mean absolute error (MAE), defined for absolute LOP over an m-country sample as:

$$\text{MAE} = \left(\sum_{j=1}^{m} \left| \text{RPL}_j^A - 1 \right| \right) / m \tag{10}$$

Similar computations may be made for relative RPL and LOP, and the summation may run over time periods as well as countries.

Because the norm ("true") value of RPL is unity, the mean absolute percentage error (MAPE)--meaning percentage of the true value--is given simply by

$$MAPE = 100 \cdot MAE \tag{11}$$

The variance around the mean is misleading, as the mean RPL may differ from the norm value. Rather, <u>a second appropriate measure is the root-mean-squared percentage error (RMSPE)</u>, analogous to the variance in that it penalizes deviations of the RPL from unity not linearly (as does the MAE or MAPE) but according to a quadratic loss function:

$$RMSPE = 100 \ \sqrt{\underset{j=1}{\overset{m}{R}} \ (RPL_j^A - 1)^2/m} \tag{12}$$

The RMSPE is the mean-squared percentage error corrected for the squaring via the over all square root, thereby making the statistic comparable in level to the MAPE. Of course, if the scalar 100 is dropped, the conventional root-mean-squared error (RMSE) is obtained.

5. Researchers that undertake direct computation of absolute RPL have the problem of obtaining comparable data across countries. The safest solution is to obtain primary-product prices from the international commodity exchanges [Jain (1980), Protopapadakis and Stoll (1983, 1986)] or to construct one's own price indices from first principles [Kravis and Lipsey (1971)]. In either case, only a handful of countries can enter the sample. Indeed, no study appearing in Table 2.1 considers more than five countries (see the second column). <u>A rigorous testing of absolute LOP demands a large sample of countries</u>.

4.4.1. Data for New Investigation of the LOP

The International Comparison Project (ICP) provides Geary-Khamis absolute PPP and RPL data on a disaggregate, national-accounts expenditure basis. The completed Phase IV [United Nations (1987)] covers sixty countries for 1980, and data for twenty-two countries for 1985 are thus far available in Phase V [OECD (1988)]. Although ICP data have been used to test the absolute LOP at aggregate (GDP and tradable/nontradable) levels--a practice continued in Section 5 of this paper--these data are ignored by investigators of the disaggregate LOP. Indeed, absorption as distinct from output data are not used at all (except insofar as export and import prices and unit values can be described as such, see Table 1). Yet the ICP under the rubric of the United Nations will be the principal source of absolute PPP data for years to come, and therefore it behooves researchers of the LOP to maximize use of ICP output.

4.5. Results with ICP Data

The ICP exhibits a disaggregation of the "food, beverages, and tobacco" category of consumption into ten components, and any disaggregation at so fine a level is unusual for national-accounts statistics. With the n-commodity set constituting the entirety of GDP, Geary-Khamis PPP data for the ten components are obtained as s-commodity subsets (see Section 2.2.1) via equations (5) and (6) and RPL computed via equation (8). All these computations are done by the ICP, though subject to its own peculiarities.[11] The United States is the numeraire country.

The MAPE and RMSPE for each of the ten commodity groups are presented in Table 2.2. For 1980, the MAPE (RMSPE) is in the range of 30-38 (38-47) percent for all categories except "alcoholic beverages" and "tobacco." For 1985, "fish," "milk, cheese, and eggs," and "other food" (coffee, tea, cocoa, spices, sweets, sugar) exhibit smaller deviations under either criterion than do the other items. Further, deviations are uniformly smaller--irrespective of category or criterion--in 1985 than in 1980. This result is to be expected, as the 1985 sample is much more homogeneous, consisting solely of OECD countries. With greater similarity of economies, the LOP is naturally closer to fulfillment (see Section 4.1.3).

Nevertheless, with no MAPE or RMSPE even below 20 percent and reaching as high as 126 percent, must the LOP be rejected out of hand? The answer is negative, because special factors are at work to widen divergences from the LOP for the commodity groups examined. First, with the PPP data on a national-accounts expenditure basis, the agricultural prices being compared are domestic market prices rather than auction prices on organized commodity exchanges; but it is the latter that provide conditions among the most favorable for the LOP.[12]

Second, "alcoholic beverages" and "tobacco"--the items with the largest deviations--are subject to special indirect taxes in many countries and to monopolistic (nationalized) distribution in some countries. Third, differential price support schemes act to segment markets of agricultural products in the various countries.

Fourth, although the products listed in Table 2.2 are conventionally considered "tradables," in fact they embody a "nontradable" (services)

TABLE 2.2. Absolute Real Price Levels: Mean Percentage
Errors

Commodity Group	Absolute (MAPE)		Root-Squared (RMSPE)	
	1980[a]	1985[b]	1980[a]	1985[b]
Bread and cereals	32	31	38	35
Meat	38	29	47	42
Fish	35	23	44	28
Milk, cheese, and eggs	30	21	38	25
Oils and fats	31	31	41	34
Fruits and vegetables	31	30	40	34
Other food	32	22	38	25
Non-alcoholic beverages	32	30	45	35
Alcoholic beverages	81	39	126	45
Tobacco	41	35	62	43

DATA SOURCES:1980--United Nations (1987, pp. 35-40), 1985--
OECD (1988, pp. 26-27).

[a]Number of observations: Non-alcoholic beverages, tobacco-
-60; alcoholic beverages--59 (excludes Pakistan); remaining
items--59 (excludes Finland).

[b]Number of observations: 22.

component. The reason is the ICP adoption of a final-expenditure rather than industry-of-origin breakdown of GDP. As Maddison (1983, p. 34) notes: "Many activities which are 'services' from a production perspective are 'disguised' in the . . . [ICP] approach because they are embedded in the end-use product. Thus, consumption of rice, meat, fish, milk, and so on are 'commodity' activities valued at retail marketprices, whereas the production approach measures the agricultural and the trading [that is, distributive-services] components separately at factor cost." "Nontradables," such as the distributive services included in the market prices of agricultural commodities, are less susceptible to international arbitrage than are tradables, and reliance must be placed on similarities of economies for adherence to the LOP. It is probable, however, that the "distributive margin (for transport, wholesale, and retail trade)" is higher for higher-income or more-developed countries, acting to widen divergences from the LOP.[13]

5. THE LAW OF ONE PRICE AT AN AGGREGATE LEVEL

5.1. Traditional Approach

5.1.1. Definitions of Variables

Turning now to the absolute LOP at the highest level of aggregation, GDP itself with PPP_j^A at the GDP level, the RPL is called the "national price level" (denoted as PL_j for country j), as noted in Section 2.3, and is defined as:

$$PL_j = PPP_j^A /E_j \tag{13}$$

In the 1980s a methodologically uniform econometric literature developed to explain PL.[14] It takes for granted that the LOP does not hold at the level of GDP ($PL_j \neq 1$) and seeks not directly to test the LOP for PL but rather to explain the value of PL across countries.

Letting Y_{jk} denote the GDP of country j valued at prices of country k (called "real income" when j varies with k fixed) and u denote the numeraire country, then

$$Y_{ju} = Y_{jj}/PPP_j \tag{14}$$

A fundamental abstraction of the literature, retained in this paper, is the division of output into two sharply defined classes: tradables and nontradables (sometimes loosely called commodities and services).[15] Letting YT_j (YN_j) denote country j's output of tradables (nontradables) valued at country j's prices,

$$Y_{jj} = YT_j + YN_j \tag{15}$$

Correspondingly, the share of tradables (nontradables), ST_j (SN_j), in country j's output is defined as

$$ST_j = YT_j/Y_{jj} \qquad SN_j = YN_j/Y_{jj} \tag{16}$$

where, of course,

$$ST_j = 1 - SN_j \tag{17}$$

Further, with PT_j (PN_j) denoting the ratio of the price level of tradables (nontradables) in country j to that in the numeraire country, then country j's nontradable/tradable price-level ratio, P_j, is defined as

$$P_j = PN_j/PT_j \tag{18}$$

5.1.2. Deficiencies

The existing literature exhibits several deficiencies that will be remedied in the present study. First, the literature ignores the fact that there exists an analytical relationship between the national price level, PL, and the nontradable/tradable price-level ratio, P, and that this relationship depends on the PPP index adopted. Instead, authors simply use as the principal explanatory variables for PL those country characteristics (for example, per capita real income) that they deem to affect P.[16]

Second, to validate this procedure, the LOP for tradables is imposed, that is,[17]

$$PT_j = E_j \tag{19}$$

In doing so, explanatory variables for PL emanating from nonfulfillment of the LOP are excluded.[18]

Third, even though fulfillment of the LOP at the GDP level means definitionally a unitary PL, authors reject unity as the norm value of

PL.[19]   While they are correct in avoiding routine acceptance of the LOP at the GDP level (though paradoxically, as noted above, they adopt strict LOP for tradables), they take the additional, unwarranted, step of adopting the actual value of PL as its norm value; for the objective of their econometrics is to minimize the sum of squared residuals from the observed values of PL. In so doing, in accepting the LOP for tradables and in rejecting unity as the norm value of PL, empirical work loses its focus: researchers fail to search explicitly for reasons why PL differs from unity--why the LOP is not fulfilled at the tradable and GDP levels--and to select explanatory variables for PL on this basis.

5.2. An Alternative Approach

5.2.1. Relationship Between PL and P

An approach to explaining the national price level that avoids these deficiencies must begin by deriving the analytical relationship between PL and P. As ICP data will be used, the specific PPP index adopted is again Geary-Khamis.[20]

Following the practice of the ICP [see, for example, KHS (1982, p. 90)], international prices are defined relative to numeraire-country prices. Let PT_I (PN_I) denote the ratio of the international price level of tradables (nontradables) to the price level of tradables (nontradables) in the numeraire country. Recalling from Section IB1 that the Geary-Khamis PPP is a Paasche index except that the "world" replaces the base country, and re-expressing the index in terms of price relatives, with the tradable/nontradable GDP breakdown, equation (2) reduces to:

$$PPP_j^A = \frac{1}{\sum_j ST_j \cdot (PT_I/PT_j) + SN_j \cdot (PN_I/PN_j)} \qquad (20)$$

While the inverse of (PT_I/PT_j) or (PN_I/PN_j) in equation (20) has dimension "number of units of currency j per unit of 'international' currency," as befits a Paasche index; the dimension of PPP_j^A remains "number of units of currency j per unit of numeraire (rather than only 'international') currency," for PPP_u is set equal to unity (recalling that PPP is at the GDP level; see Section 2.2.1).

The international/numeraire-country price level of tradables (nontradables) is defined as the ratio of world output at international prices to world output at numeraire-country prices:

$$PT_I = \frac{\underset{j \neq u}{R} \; YT_j/PPP_j^A \; + \; YT_u}{\underset{j \neq u}{R} \; YT_j/PT_j \; + \; YT_u} \qquad (21)$$

$$PN = \frac{\underset{j \neq u}{R} \; YN_j/PPP_j^A \; + \; YN_u}{\underset{j \neq u}{R} \; YN_j/PN_j \; + \; YN_u} \qquad (22)$$

There is an asymmetry in that there is only one international-currency/j-currency conversion factor $(1/PPP_j^A)$ and it pertains to all output (as does the international-currency/numeraire-currency conversion factor, $1/PPP_u^A = u$), whereas the numeraire-currency/j-currency conversion factor $(1/PT_j$ or $1/PN_j)$ is specific to tradables or nontradables.

With m countries, there are $(m - 1)$ equations (20). Substituting for PT_I and PN_I via equations (21) and (22) yields a system of $(m - 1)$ nonlinear equations determining the PPP_j^A, $j \neq u$, the analytical solution of which would be difficult indeed and is not attempted for a general m. Instead, the specific case of m = 2 is considered, in which case the summation signs in (21) and (22) vanish. Applying (13), after considerable involved algebra, one obtains

$$PL_j = (PT_j/E_j) \; K \; \frac{-A + \sqrt{A^2 + 4 KST_u KST_j KSN_u KSN_j KP_j KY_{ju}/Y_{uu}}}{2 KST_j KSN_j KY_{ju}/Y_{uu}} \qquad (23)$$

where $A = ST_u KST_j KSN_u KP_j + ST_u KSN_u KSN_j$

$$- ST_j KSN_u KSN_j KP_j KY_{ju}/Y_{uu} - ST_u KST_j KSN_j KY_{ju}/Y_{uu}$$

5.2.2. Justification for Norm Value of Unity

Having derived the PL-P analytical relationship (23), it will now be demonstrated that the norm value of PL is unity. Consider the LOP for tradables, equation (19), combined with an identical tradable/nontradable price structure in country j and the numeraire country, that is, a unitary P_j. If $PT_j = E_j$ and $P_j = 1$, then PL_j reduces to unity, as these substitutions into (23) yield after involved algebra.

Thus the LOP for tradables and an identical tradable/nontradable price structure across countries imply a unitary PL.[21]

Appropriate econometric methodology is now apparent. There are two sets of forces that move PL away from its norm value of unity (in either direction)--first, those factors that cause PT/E to deviate from unity; and second, those elements that make P diverge from unity. The national price level is expressed as the absolute deviation from its norm value (unity), and explanatory variables operate by taking either PT/E or P, away from unity.

5.3. Variables for Hypothesis Testing

Applying the new approach to the national price level econometrically, the dependent variable is the absolute value of the percentage deviation of country j's national price level (PL_j) from unity, denoted as $PLDEV_j$:

$$PLDEV_j = 100 \cdot |PL_j - 1| \qquad (24)$$

The first group of explanatory variables for PLDEV are the forces that give rise to deviations from the LOP for tradables, that is, make PT/E diverge from unity. These forces, discussed in Section 4.1.2., are not measured directly in empirical studies of the LOP even at disaggregate levels. At the far more challenging aggregate level, one must resort to ad hoc variables, and these capture only two of the many elements mentioned in Section 4.1.2.[22] First, firms with monopoly power may practice price discrimination between domestic and foreign markets. An admittedly crude proxy for the lack of monopoly power of the firms in a country's tradable sector is the country's real GDP (on the grounds that the bigger the domestic market, the greater the scope for competition). Defining $YREL_j$ as the real GDP of country j relative to the numeraire country:

$$YREL_j = 100 \cdot (Y_{ju}/Y_{uu}) \qquad (25)$$

this variable has a theoretically negative effect on $PLDEV_j$ (meaning that a larger YREL implies a smaller PLDEV).

Second, oligopolists may absorb the impact of a changing exchange rate in their profits, so that the price of tradables does not move with the exchange rate to maintain the LOP. Greater variability in the exchange rate enhances this phenomenon. The coefficient of variation (ratio of standard deviation to mean) of the exchange rate (number of units of domestic currency per SDR--this multilateral measure of foreign exchange deemed superior to a single currency), $RVAR_j$, ideally would determine $PLDEV_j$ interactively with the measure of market power, $1/YREL_j$, in the form of the variable

$$RVARY_j = RVAR_j/YREL_j \qquad (26)$$

with a theoretically positive effect on PLDEV. However, given the crudeness of the measure of market power, it could be that RVAR is a variable superior to RVARY (on the grounds that assuming equal market power of countries' tradable sectors is superior to proxying this market power by the inverse of real GDP).

The second group of explanatory variables for PLDEV are the factors that cause P to diverge from unity. Recalling that P_j is the nontradable/tradable price structure in country j relative to that in the numeraire country, its determinants are the usual general-equilibrium variables--demand conditions, technology, factor supplies, market structure, and government policies--in country j relative to the numeraire country. These are just the elements determining the similarity of economies (see section 4.1.3). As proxy general-equilibrium variables, per capita real income (YC_{ju} for country j) and the share of nontradables (SN_j) are used. Specifically, the income variable, $YCDEV_j$, is the absolute value of the percentage deviation of country j's per capita real GDP from that of the numeraire country:

$$YCDEV_j = 100 \cdot |(YC_{ju} - YC_{uu})/YC_{uu}| \qquad (27)$$

and the share of nontradables enters as the absolute value of the deviation of country j's share of nontradables (in percent) from that of the numeraire country, $SNDEV_j$:

$$SNDEV_j = 100 \cdot |SN_j - SN_u| \qquad (28)$$

Both variables are expected to have a positive effect on $PLDEV_j$. The greater the dissimilarity between country j and the numeraire country, the greater is the deviation of P_j from unity and therefore the greater is $PLDEV_j$.

5.4. The Samples

The ideal sample would involve price (PL) and output (Y_{ju}, YC_{ju}, SN_j) variables for the year 1980 from Phase 4 of the ICP--the sixty-country sample used in sections 4.4, 4.5. The problem is that, for ICP

Phases 4 and 5, data are lacking to divide GDP into tradables (YT) and nontradables (YN) on a final-expenditure basis, and therefore to construct the variable SN and then SNDEV on that basis, consistent with the final-expenditure breakdown of PPP (see section 4.5). Only for Phases 1-3 are such data available. Because Phase 3 [KHS (1982)] has 34 observations, more than double the sample size for Phases 1 and 2 [KKHS (1975) and KHS (1978)], it provides the primary sample; the year is 1975. Tradables are defined as final expenditure on commodities except construction, with nontradables final expenditure on services and construction [KHS (1982, pp. 69, 191-93)].

Nevertheless, in view of its size and use in investigating the LOP at the disaggregate level in Section 4.5, the sixty-country (1980) sample of Phase 4 warrants inclusion as a second sample. Therefore, to construct SN and then SNDEV for the 1980 sample, resort is made to an industry-of-origin specification. Tradable production (YT) is defined as output originating in (i) agriculture, hunting, forestry, and fishing, (ii) mining and quarrying, and (iii) manufacturing; nontradable production (YN) as the remainder of GDP at factor cost (output arising in construction and services-producing sectors). The resulting SN variable is inferior to that developed on a final-expenditure basis because market prices of tradable commodities (such as the products considered in sections 4.4, 4.5) are inclusive of distributive services, whereas the industry-of-origin division of tradables/nontradables places these services in nontradables. Because the distributive margin of market prices is likely to vary positively with per capita real income or level of development, an "errors in variable" is generated for SNDEV founded on the industry-of-origin approach to tradables/ nontradables.[23]

After excluding centrally planned economies, because of inapplicable hypotheses and lack of data, and Guatemala (for 1980) because of lack of data to construct SN, the samples are reduced to 31 countries for 1975 and 57 for 1980. The numeraire country is the United States: PPP and real output are based on the Geary-Khamis index. Data sources and construction of variables are described in the Appendix.

5.5. Results

In his conventional econometric investigation of PL using Phase-4 data, Clague (1986, p. 20) notes that African countries have "remarkably high [national] price levels considering their very low real incomes," and that this is also true to a lesser extent for European countries. To capture regional influences on PL not covered by the continuous explanatory variables, the dichotomous variables AFRDUM (1 if African country, 0 otherwise) and EURDUM (similarly defined) are included in the regression for each year. Table 2.3 presents the two regressions; variables with coefficients having a t-value below unity (that is, variables making a negative contribution to R^2 of .86 and the explanatory variables YCDEV, SNDEV, and RVAR all significant at the 5 percent level or better. The explanatory power of the 1980 regression, in contrast, is much lower. Its main feature is the inclusion of AFRDUM with a coefficient that is both large (over 170 percent the magnitude of the general constant) and significant (at the 1 percent level). Clague suggests close monetary ties of African countries with Europe as the explanation for the similar impact of this variable in his conventional equations for PL. However, the present study indicates that the problem <u>may be</u> one of biased data, not of monetary arrangements. Consider the interaction of real GDP and the African dummy:

$$AFRY = AFRDUM \cdot YREL$$

This variable, included in the 1980 regression, is just shy of significance at the 5 percent level, and its coefficient swamps that of YREL in magnitude. In contrast, interactions of AFRDUM with the other continuous variables (YCDEV, SNDEV, and RVAR) all make negative contributions to R^2. Recalling that YREL is a direct measure of real output, these findings suggest that the Phase-4 ICP data (real price levels and output) may be inherently out of line for Africa compared to other regions in the sample. The problem could be one of incorrect data rather than hypothesis specification that excludes monetary arrangements. If so, the results for disaggregate LOP in section 4.5 could be affected should the problem be inclusive of the "food, beverages, and tobacco" component of final-expenditure GDP.

Variables measuring dissimilarity of economies (YCDEV and SNDEV) are highly significant in both regressions, whereas variables representing forces causing divergences from the LOP for tradables

TABLE 2.3. PLDEV Regression Equations

Year	1975	1980
Constant	-2.61 (0.72)	12.61 (2.86)
YDEV	0.28** (3.64)	0.24* (2.65)
SNDEV	1.37** (5.06)	0.90** (3.03)
YREL		-0.14 (1.16)
RVAR	1.52* (2.14)	-0.53 (1.43)
AFRDUM		-21.78** (4.64)
AFRY		-8.95 (1.98)
No. Obs.	31	57
$\bar{R}^2$	.86	.56

*Significant at 5 percent level.
**Significant at 1 percent level.

(YREL, RVAR, and RVARY) are less significant. Even granted the ad hoc nature of all explanatory variables, the importance of similarity among economies in influencing the LOP is clearly seen.

6. PRINCIPAL CONCLUSIONS

1. While many obstacles to perfect commodity arbitrage and perfect homogeneity of commodities exist, thereby detracting from the LOP, similarity of economies is a powerful force operating in favor of the LOP.
2. Investigators of the LOP are advised to concentrate on absolute rather than relative LOP and to utilize the ongoing data provided by the ICP.
3. Segmentation of the investigations of LOP at the aggregate and disaggregate levels should be ended.
4. Phase-4 ICP data for African countries may be flawed.

Appendix: The Data

PL:1975--computed as PPP/E; PPP: KHS (1982, pp. 176-79), E: KHS (1982, p. 10). 1980--United Nations (1987, pp. viii-ix).

Y_{ju}:1975--KHS (1982, p. 10). 1980--computed as product of YC_{ju} and population (except that direct figure is given for the United States), from United Nations (1987, pp. viii-ix).

YC_{ju}:1975--KHS (1982, p. 12). 1980--United Nations (1987, pp. viii-ix).

RVAR:Computed from 13 end-of-month observations, December 1974 (1979) to December 1975 (1980). Sources: International Monetary Fund (various issues) and Cowitt (1985), the latter for Hong Kong and Zimbabwe. Domestic-currency/SDR rate where not provided directly, is computed as product of domestic-currency/dollar and dollar/SDR rates.

SN:Computed as YN/(YT + YN), with YT and YN obtained as follows: 1975 (expenditure basis)--KHS (1982, p. 194). 1980 (industry-of-origin basis)--as described in Section 5.4, with most data from United Nations (1986a). Data for Yugoslavia are from OECD (1985). For Madagascar, construction (a nontradable flow) is separated from "mining and quarrying, and manufacturing" on the basis of the 1974 ratio [from United Nations (1985)], but "electricity, gas, and water" is unavoidably included in tradable output, as it is for Senegal (1980). For Spain, "electricity, gas, and water" is separated from "mining and quarrying, and manufacturing" on the basis of the 1977 ratio. For Malawi and the Ivory Coast, and for Mali, data used are for 1979 and 1981, respectively.

NOTES

1. For discussions of these three and related indices in the context of PPP, see Ruggles (1967), Kravis et al. (1975, chs. 4-5), KHS (1978, ch. 3), KHS (1982, ch. 3), Diewert (1987), and Officer (1989). Diewert's exposition is extremely technical.

2. See KKHS (1975, pp. 68-69).

3. Replacing PPP_j^A by PI_j and letting j and b on the right-hand sides of equations (1) and (2) represent periods t and 0, respectively, these equations define the Laspeyres and Paasche PI_j for a given country j.

4. The term "national price level" apparently originated with Kravis and Lipsey (1983, p. 9). The nomenclature "real price level" was suggested by Clague (1985, p. 998) because it is a "ratio of two prices" rather than a money (nominal?) price level ("the price of goods in terms of money"). A second reason to adopt Clague's terminology is that the concept "real exchange rate" as used in the theoretical literature is the exchange rate corrected for absolute PPP: the RPL is the reciprocal of the real exchange rate. In the empirical literature, in contrast, the real exchange rate pertains to relative PPP and is the reciprocal of relative RPL (defined below). See Kravis and Lipsey (1983, p. 9, esp. n. 3). Going beyond Clague, it is only logical to extend the term "real exchange rate" to disaggregate price levels instead of confining it to total output (GDP), and to delineate absolute and relative RPL in concordance with absolute and relative PPP.

5. "'Pure competition' is . . . competition unalloyed with monopoly elements"--Chamberlin (1958, p. 6).

6. "'Perfect' competition . . . may imply . . . an absence of friction in the sense of an ideal fluidity or mobility of factors such that adjustments to changing conditions which actually involve time are accomplished instantaneously in theory . . . perfect knowledge of the future and the consequent absence of uncertainty"--Chamberlin (1958, p. 6).

7. Dornbusch (1987, pp. 95, 98-102) presents various oligopolistic models with differentiated products that generate varying divergences from the LOP when there is an exchange-rate change.

8. Roll (1979) develops a model of the LOP in an efficient-markets framework. A good summary and assessment of it is provided by Shapiro (1983, pp. 301-305). A more-rudimentary efficient-markets approach to the LOP is in Magee (1978, pp. 69-70).

9. Bordo and Choudhry (1977) and Magee (1979) supplement this procedure with formal time-series analysis.

10. See Protopapadakis and Stoll (1986), the only justifiable study in group V, and note the discussions in CCM (1982, pp. 331-32) and Officer (1986, p. 163).

11. See OECD (1988, pp. 4-9) and United Nations (1987, pp. 1-14).

12. This is the view of Isard (1974, p. 379; 1977, p. 942), Jain (1980, p. 89), Protopapadakis and Stoll (1983, p. 1433), and Dornbusch (1987, p. 1079).

13. See Maddison (1983, p. 37, n. 12) and Kravis (1986, pp. 22-23).

14. Highlighting this literature are the studies of Clague (1986, 1988a, 1988b), and Kravis and Lipsey (1983, 1987). Outside the mainstream and in opposition to it is the work of Officer (1989).

15. As Clague (1988a, p. 238) writes: "It is obvious, moreover, that the tradability of goods is a matter of degree rather than a simple dichotomy, but theorists need to make simplifications, and the assumption of a tradable/nontradable dichotomy has proven quite useful in thinking about national price levels." While theoretical work on PL on rare occasion divides tradables into exportables and importables [Clague (1985)] or even treats the tradability characteristic of commodities as a continuum [Houthakker (1978), Marris (1984)], the econometric literature relies solely on the tradable/nontradable distinction.

16. For example, Clague (1986, p. 321) states: "the key to the change in RPL [the national price level] is found by looking at the relative prices of services and commodities."

17. Thus, Clague (1986, p. 321) notes that his "theoretical model . . . assumes that commodity prices are equalized through trade while service prices are not."

18. Clague (1986, 1988a) and Kravis and Lipsey (1983) do include a money-growth variable in the PL equation, thereby recognizing deviations from the law of one price via Dornbusch's (1976) "overshooting" hypothesis. However, the variable used is short-run and monetary rather than long-run and structural (a "country characteristic"), contradictory not only to the other explanatory variables for PL but also to the forces accounting for divergences from the LOP (see Sections 4.1.2, 4.1.3). Further, simply incorporating the money growth rate as an additive term in the PL equation is inconsistent modeling--adopting the LOP for hypothesis specification via P, thus obtaining the country-characteristic variables, and then abandoning the LOP for inclusion of the money-growth variable. For a detailed critique of this and other aspects of the existing literature, see Officer (1989).

19. For example, Clague (1986, p. 320) states: "The idea that in long-run equilibrium PPPs tend to equal exchange rates has been rather conclusively refuted by the data." Kravis and Lipsey (1987, p. 97) write: "Equality of price levels--sometimes referred to as the 'law of one price' or 'purchasing power parity'--is not the norm."

20. The, unconventional, formulation of the Geary-Khamis index to follow is consistent with the descriptions in KKHS (1975, pp. 68-70) and KHS (1978, pp. 73-76; 1982, pp. 89-90). For derivations of the analytical relationship between PL and P for other PPP indices (including the Laspeyres and Paasche), see Officer (1989).

21. The statement has been proved for $m = 2$. The equivalent of equation (23) for general m would be complicated indeed, with PL_j for a given j depending on SN_j, P_j, and Y_{ju} for all j. Intuition, though, would support the proposition even in this case.

22. An extremely important source of divergences from the LOP is product differentiation of a country's tradables, but there exists no acceptable measure of this characteristic at the aggregate level. While the literature on intra-industry trade uses the commodity diversification of exports for this purpose [see Havrylyshyn and Civan (1983, p. 121)], the limitations of this proxy are overwhelming. One problem is that the product differentiation of exports may differ from that of all tradables because of the sensitivity of exports to demand conditions abroad.   A more serious defect is that commodity concentration is not antithetical to product differentiation. Production

of tradables concentrated in a commodity sense might nevertheless by highly differentiated in the commodities produced.

23. It is not contradictory that there exists abundant evidence that, on its own terms, the industry-of-origin tradable/nontradable breakdown is empirically sound. See Officer (1986, pp. 176-78) and Kravis and Lipsey (1987, p. 128, n. 8).

REFERENCES

Bigman, D. (1983). "Exchange Rate Determination: Some Old Myths and New Paradigms." in D. Bigman and T. Taya (eds.), Exchange Rate and Trade Instability Cambridge, MA: Ballinger Publishing Company.

Bordo, M.D., and E. Choudhry (1977). The Behavior of the Prices of Traded and Nontraded Goods: The Canadian Case, 1962-74. Carleton University Discussion Paper 77-01, Ottawa, Canada.

Chamberlin, E.H. (1958). The Theory of Monopolistic Competition. Cambridge, MA: Harvard University Press.

Clague, C. (1985). "A Model of Real National Price Levels." Southern Economic Journal 51 (April): 998-1017.

_____ (1986). "Determinants of the National Price Level: Some Empirical Results,"Review of Economics and Statistics 58 (May): 320-23.

_____ (1988a). "Explanations of National Price Levels," in J. Salazar-Carrillo and D.S. Prasada Rao (eds.), Word Comparison of Incomes, Prices and Product. Amsterdam: North-Holland.

_____ (1988b). "Purchasing-Power Parities and Exchange Rates in Latin America." Economic Development and Cultural Change 36 (April): 529-41.

Cowitt, P.P. (1985). 1984 World Currency Yearbook. Brooklyn, NY: International Currency Analysis.

Crouhy-Veyrac, L., M. Crouhy, and J. Melitz (1982). "More About the Law of One Price." European Economic Review 18 (July): 325-44.

Curtis, J.M. (1971). "Direct Foreign Influences on Canadian Prices and Wages," in R.E. Caves and G.L. Reuber (eds.), Capital Transfers and Economic Policy: Canada, 1951-1962. Cambridge, MA: Harvard University Press.

Daniel, B.C. (1986). "Empirical Determinants of Purchasing Power Parity Deviations." Journal of International Economics 21 (November): 313-26.

Diewert, W.E. (1987). "Index Numbers." in J. Eatwell, M. Milgate, and P. Newman (eds.), The New Palgrave, vol. 3. London: Macmillan.

Dornbusch, R. (1976). "Expectations and Exchange Rate Dynamics." Journal of Political Economy 84 (December): 1161-76.

_____ (1987). "Purchasing Power Parity." in J. Eatwell, M. Milgate, and P. Newman (eds.), The New Palgrave, vol. 3. London: Macmillan.

Dunn, R.M., Jr. (1970). "Flexible Exchange Rates and Oligopoly Pricing: A Study of Canadian Markets." Journal of Political Economy 78 (January/February): 140-51.

_____ (1973). "Flexible Exchange Rates and Traded Goods Prices: The Role of Oligopoly Pricing in the Canadian Experience," in H.G. Johnson and A.K. Swoboda (eds.), The Economics of Common Currencies. Cambridge, MA: Harvard University Press.

Fromm, G. (1974). "Comment." in P.B. Clark, D.E. Logue, and R.J. Sweeney (eds.), The Effects of Exchange Rate Adjustments. Washington, DC: Dept. of the Treasury.

Genberg. H. (1975). World Inflation and the Small Open Economy. Stockholm: Swedish Industrial Publications.

_____ (1978). "Purchasing Power Parity under Fixed and Flexible Exchange Rates." Journal of International Economics 8 (May): 247-76.

Goodwin. B.K. (1988). Empirically Testing the Law of One Price in International Commodity Markets: A Rational Expectations Approach. Ph.D. diss., Dept. of Economics and Business, North Carolina State University, Raleigh.

Havrylyshyn, O., and E. Civan (1983). "Intra-Industry Trade and the Stage of Development: A Regression Analysis of Industrial and Developing Countries," in P.K.M. Tharakan (ed.), Intra-Industry Trade. Amsterdam: North-Holland.

Houthakker, H.S. (1978). "Purchasing Power Parity as an Approximation to the Equilibrium Exchange Ratio." Economics Letters 1: 71-75.

International Monetary Fund (various issues). International Financial Statistics, Washington, D.C.

Isard, P. (1974). "The Price Effects of Exchange-Rate Changes." in P.B. Clark, D.E. Logue, and R.J. Sweeney (eds.), The Effects of Exchange Rate Adjustments. Washington, DC: Dept. of the Treasury.

_____ (1977). "How Far Can We Push the 'Law of One Price'?" American Economic Review 67 (December): 942-48.

Jain, A.K. (1980). Commodity Futures Markets and the Law of One Price. Michigan International Business Studies Number 16. Graduate School of Business Administration, University of Michigan, Ann Arbor.

Johnson, P.R., T. Grennes, and M. Thursby (1979). "Trade Models with Differentiated Products." American Journal of Agricultural Economics 61 (February): 120-27.

Kravis, I.B. (1986). "The Three Faces of the International Comparison Project." The World Bank Research Observer 1 (January): 3-26.

Kravis, I.B. and R.E. Lipsey (1971). Price Competitiveness in World Trade. New York: Columbia University Press.

_____ (1974). "International Trade Prices and Price Proxies," in Nancy D. Ruggles (ed.), The Role of the Computer in Economic and Social Research in Latin America. New York: Columbia University Press.

_____ (1977). "Export Prices and the Transmission of Inflation." American Economic Review 67 (February): 155-63.

_____ (1978). "Price Behavior in the Light of Balance of Payments Theories." Journal of International Economics 8 (May): 193-246.

_____ (1983). Toward an Explanation of National Price Levels. Princeton Studies in International Finance, No. 52. Princeton: Princeton University Press.

_____ (1987). "The Assessment of National Price levels." in S.W. Arndt and J.D. Richardson (eds.), Real-Financial Linkages among Open Economies. Cambridge, MA: MIT Press.

Kravis, I.B., A. Heston, and R. Summers (1978). International Comparisons of Real Product and Purchasing Power. Baltimore: Johns Hopkins University Press.

_____ (1982). World Product and Income. Baltimore: Johns Hopkins University Press.

Kravis, I.B., Z. Kenessey, A. Heston, and R. Summers (1975). A System of International Comparisons of Gross Product and Purchasing Power. Baltimore: Johns Hopkins University Press.

Maddison, A. (1983). "A Comparison of the levels of GDP Per Capita in Developed and Developing Countries, 1700-1980." Journal of Economic History 43 (March): 27-41.

Magee, S.P. (1978). "Contracting and Spurious Deviations from Purchasing-Power Parity," in J.A. Frenkel and H.G. Johnson (eds.), The Economics of Exchange Rates. Reading, MA: Addison-Wesley.

_____ (1979). "A Two-Parameter Purchasing-Power-Parity Measure of Arbitrage in International Goods Markets," in J.P. Martin and A. Smith (eds.), Trade and Payments Adjustment under Flexible Exchange Rates. London: Macmillan.

Marris, R. (1984). "Comparing the Income of Nations." Journal of Economic Literature 22 (March): 40-57.

Norman, N.R. (1978). "On the Relationship Between Prices of Home-Produced and Foreign Commodities." Oxford Economic Papers 27 (November): 426-39.

Organization for Economic Cooperation and Development (1985). National Accounts 1971-1983, vol. II Paris: OECD.

_____ (1988). Purchasing Power Parities and Real Expenditures 1985. Paris: OECD.

Office of Management and Budget (1987). Standard Industrial Classification Manual 1987, Washington, D.C.

Officer, L.H. (1986). "The Law of One Price Cannot Be Rejected: Two Tests Based on the Tradable/Nontradable Price Ratio." Journal of Macroeconomics 8 (Spring): 159-82.

_____ (1989). "The National Price Level: Theory and Estimation." Journal of Macroeconomics 11 (Summer).

Ormerod, P. (1980). "Manufactured Export Prices in the United Kingdom and the 'Law of One Price'." Manchester School of Economic and Social Studies 48 (September): 265-83.

Protopapadakis, A., and H.R. Stoll (1983). "Spot and Futures Prices and the Law of One Price." Journal of Finance 38 (December): 1431-55.

_____ (1986). "The Law of One Price in International Commodity Markets: A Reformulation and Some Formal Tests." Journal of International Money and Finance 5 (September): 335-60.

Richardson, J.D. (1978). "Some Empirical Evidence on Commodity Arbitrage and the Law of One Price." Journal of International Economics 8 (May): 341-51.

Roll, R. (1979) "Violations of Purchasing Power Parity and Their Implications for Efficient International Commodity Markets," in M. Sarnat and G.P. Szego (eds.), International Finance and Trade, vol. I. Cambridge, MA: Ballinger.

Rosenberg, L.C. (1974). "Impact of the Smithsonian and February 1973 Devaluations on Imports: A Case Study of Steel," in P.B. Clark, D.E. Logue, and R.J. Sweeney (eds.), The Effects of Exchange Rate Adjustments, Washington, DC: Dept. of the Treasury.

Ruggles, R. (1967). "Price Indexes and International Price Comparisons," in Ten Economic Studies in the tradition of Irving Fisher, New York: John Wiley.

Shapiro, A.C. (1983). "What Does Purchasing Power Parity Mean?" Journal of International Money and Finance 2 (December): 295-318.

Smith, H.H. (1988). "U.S. Agricultural Export Competitiveness: Export levels, Trade Shares, and the Law of One Price." Federal Reserve Bank of Dallas Economic Review (July): 14-25.

Solnik, B. (1979). "International Parity Conditions and Exchange Risk," in M. Sarnat and G.P. Szego (eds.), International Finance and Trade, vol. I. Cambridge, MA: Ballinger.

U.S. Bureau of Labor Statistics (1987). Producer Price Indexes, Data for December 1987, Washington, D.C.: U.S. Government Printing Office.

United Nations (1985). National Accounts Statistics: Main Aggregates and Detailed Tables, 1982. New York: United Nations.

_____ (1986a). National Accounts Statistics: Main Aggregates and Detailed Tables, 1983. New York: United Nations.

_____ (1986b). Standard International Trade Classification, Revision 3. Statistical Papers, Series M. No. 34/Rev. 3. New York: United Nations.

_____ (1987). World Comparisons of Purchasing Power and Real Product for 1980, Part Two. New York: United Nations.

Comments by Catherine L. Mann

"The Law of One Price: Two Levels of Aggregation"

by Lawrence Officer

Professor Officer takes a fresh look at the Law of One Price (LOP) in his paper, "The Law of One Price: Two Levels of Aggregation." The motivations are first, a new data set that was in essence designed to examine international LOP and second, a new methodology for testing LOP that he believes is superior to current methods. The paper has two objectives: first, to test whether LOP holds at various levels of commodity disaggregation of the data and second, to examine whether the forces causing deviations from LOP are primarily macro or micro in nature. The paper concludes that the LOP tested with aggregate data is more likely to hold between countries of similar macroeconomic characteristics such as per capita GNP. Microeconomic issues such as market structure appear to be important when testing for whether LOP holds using disaggregate data but not important when testing LOP with aggregate data.

Testing for whether the LOP holds and examining causes for its failure to hold have been examined extensively in the literature. Officer sets out to add to that literature by using disaggregated data from the Phase IV and Phase V ICP data set. When testing LOP with disaggregated data, he advocates direct testing of the absolute version of the LOP by calculating deviations of international relative prices from their norm value of unity. For testing LOP with aggregate data, he reverts to the more common econometric estimation of the determinants of the national price level.

While applauding the basic objectives of the paper--to test LOP at various levels of disaggregation and to examine causes of deviations from the LOP from a macro and micro viewpoint--I have some concerns about the implementation. First, I'm not sure that Officer achieves what he sets out to do. Examinations of whether LOP is accepted at one level of data disaggregation but rejected at different levels of data disaggregation would be more convincing if the tests were performed on exactly the same data set. While Officer uses the ICP data for both tests, he uses data from different Phases; data that purport to measure the same price movements can vary substantially from one Phase to another because of different sampling techniques. Moreover, the empirical analysis of sources of deviations from the LOP is somewhat strained; the proxies for different macro characteristics or market structure could be better chosen.

Nevertheless, the paper does contribute to the LOP literature; the large deviation from the norm value of unity for the disaggregated components of the Phase IV and V ICP data supports the conclusions of other studies that greater disaggregation does not rescue the LOP.

In my comments, I first review Officer's price definitions, since they are somewhat different from those of the standard macroeconomist; I also think some examples would help the reader understand the differences. Second, I summarize what I think are the essential issues behind the various estimation techniques and Officer's choice. Third, I discuss the empirical implementation.

Officer presents two relative price concepts. One is purchasing power parity (PPP), which looks at the relative price level or relative price index in a country as compared to a numeraire (either another country or a group of countries). The other concept is the real price level (RPL), which compares in a common currency the price level or price index of a country and a numeraire country or group of countries. (RPL is the inverse of the real exchange rate.) As for PPP, the RPL concept exists in an absolute form (comparison of levels) and a relative form (comparison of indexes over time).

To be very concrete (but also to border on the incorrect since these concepts are normally calculated for baskets of goods), the absolute PPP concept states that cars in the United States and in Germany should cost the same (in dollars and DM), since the price-specie flow mechanism would yield an exchange rate equal to one. The relative PPP concept suggests that changes in the price of cars in Germany in DM and in the United States in dollars would be the same, since macroeconomic forces in the two economies should lead to similar rates of inflation between the two countries. The RPL concepts incorporate the exchange rate into the comparison, effectively providing one more degree of freedom in the international comparison of goods-price levels and movements. The absolute RPL concept states that the DM price of cars in Germany adjusted by the dollar/DM exchange rate should equal the dollar price of cars in the United States. The relative RPL concept states that the changes in the DM prices of cars in Germany adjusted for changes in the dollar/DM exchange rate should equal the changes in the dollar price of cars in the United States.

Officer shows the essential similarity between the RPL concepts and the LOP concept as an argument for his choice of testing methodology. The LOP is the theory that the norm value of RPL should be unity--exchange rate adjusted prices of goods should be the same between countries (this is absolute RPL or absolute LOP) and differences in goods-price inflation between countries should be offset by changes in their exchange rate (this is relative RPL or relative LOP). Therefore, the appropriate test for whether the LOP holds is to examine by how much the RPL deviates from 1. Since absolute LOP is the more fundamental hypothesis, it should be tested first by examining the absolute RPL. Relative LOP is the time derivative but could hold even if the absolute LOP failed. Relative LOP should be tested by examining the deviation of relative RPL from one.

Why could RPL deviate from its norm value of one? The paper lists a host of suggestions--transport costs, trade barriers, oligopoly, product differentiation, information problems, measurement problems. Forces contributing to the LOP holding include similarities in income and economic structure between the economies, which contribute to similarities in elasticities of substitution; similarities in tastes, technology, market structures, and so forth can cause the price levels in the various economies to be similar, thus leading to RPL close to one. A good way to summarize these forces is to divide them into macroeconomic forces (levels of income, demand, spending, tastes) and microeconomic forces (market structures, product type, technology).

Officer argues that most empirical analyses of LOP do not calculate the value of RPL but instead examine whether the coefficient of a log-linear regression of price levels and exchange rates equals one. In fact, there is no distinction between these two ways of examining the hypotheses. The Officer approach first calculates RPL, then examines what forces might causes RPL to deviate from one. Using equation (10) in the paper to express the mean absolute error (MAE), assuming $m = 1$, and letting the function H be the unspecified forces causing deviation from RPL = 1, we get:

$$MAE = (RPL_j^A - 1) = H (.)$$

Using the definition of RPL_j^A, where PL is the domestic price level and PL* is the price level of the other country yields:

$$RPL - 1 = (PL/E_j^A - PL^*)/PL^*$$

Putting these together yields:

$$(PL/E)/PL^* - 1 = H(.)$$

Rearranging yields:

$$PL = E(PL^*)[1 + H(.)]$$

Taking logs yields the empirical implementation frequently used for examination of the LOP:

$$\ln (PL) = a_0 + a_1 \ln (E) + a_2 \ln (PL^*) + a_3 \ln [h(.)]$$

where $h(.)$ is a log transformation of the unspecified variables measuring forces causing deviations from LOP.

Frequently a_3 is dropped from the estimation (or, more accurately, consolidated into the constant term a_0). The test for whether the LOP holds is whether a_0 equals zero and a_1 and a_2 equal one. This test is identical to calculating MAE (the Officer approach) but in fact is more enlightening, since deviations of a_1 and a_2 from one yield clues as to where the LOP breaks down. Is the problem in differences between the two countries' price levels or is the problem in the relationship between the domestic price level and the exchange rate?

Depending on the degree of disaggregation of the data, it might make more sense to explain the deviations of a_1 and a_2 from one in terms of micro forces or macro forces. Suppose we test the LOP with disaggregated data. Macroeconomic forces such as income or demand differences might not make much difference to the prices of very disaggregated goods. But microeconomic reasons for the failure of the LOP (such as monopoly or trade barriers) might be keys in examinations using disaggregated data. Now suppose we test LOP using aggregate data. The market structure arguments for why LOP might not hold tend to wash-out across the many industries in macro aggregates, but macro arguments against LOP might be more visible in the results. So, if the LOP holds in the aggregate but not in the disaggregate, it may be because market structure arguments are important using industry-level data, but that macro forces are contributing to LOP holding at the macro level. If LOP does not hold in either disaggregate or aggregate, then at both micro and macro levels the forces against commodity arbitrage would appear to be preeminent.

Officer's first empirical effort calculates RPL for relatively disaggregated sectors using the ICP data for Phases IV and V. In general, the deviation of RPL from its norm value of unity is substantial--30 percent or more. Based on these figures, it appears that there are substantial micro forces leading to the failure of the LOP.

To test for the LOP at the macro level, Officer reverts to using an estimating equation similar to that shown above. He shows algebraically that RPL has a norm value of unity if all countries have price levels (PL) of unity. So, testing for the LOP uses cross-section data to estimate what forces might cause PL to deviate from one.

To determine what variables to put on the right-hand side, he decomposes the PL into its broad components--the exchange rate, traded goods and nontraded goods. He shows that PL has a norm value of unity if the price of traded goods equals the exchange rate (in other words, if the LOP holds for traded goods), and the relative price of traded to nontraded goods equals unity (in other words, if the traded-nontraded market structures are similar across the cross-section sample of countries). Officer's test of whether LOP holds using aggregate data depends on whether the traded-nontraded price ratio deviates from one and whether the traded prices exchange ratio deviates from one. While each of these hypotheses could be tested independently, he chooses to test them jointly.

The estimating equation has the deviation of the PL from one (the dependent variable) as a function of (1) the income relative to the numeraire country, (2) the degree of variation in its exchange rate divided by the relative income variable, (3) deviation in per capita income in the country from that in the numeraire country, (4) deviation of the share of nontraded goods in the country's economy as compared to that share in the numeraire country. All variables are cross-section "stacked" by countries in the ICP.

The first two variables are supposed to explain deviations of the traded goods price from the exchange rate. The relative income variable is supposed to proxy for monopoly power in the economy; the larger the economy, the greater the competition. The exchange rate variable is also supposed to capture market power; if an economy is oligopolistic, it should absorb more exchange rate variation into prices. However, both of these variables are macro variables intended to proxy for what is fundamentally a microeconomic concept of market structure. Why not use some information on the degree of concentration in the economies such as the share of GDP accounted for by the Fortune 500 (for the United States), the state-owned enterprises (for Africa), the remnants of the zaibatsu (for Japan), and so on.

The other two variables are supposed to proxy for why the traded-nontraded variables might deviate from unity. The deviation in the per capita income variable is based on the argument that countries with similar incomes have similar traded-nontraded structures. But this assumes no effect of development strategies (inward versus outward-oriented economies) and does not take account of differences in factor endowments that might yield different traded-nontraded structures. On the other hand, clearly the variable that measures the deviation of the share of nontraded goods in GNP between the country and the numeraire country gets at the issue of dissimilarity in trade-nontraded goods market structure between the countries.

What should we conclude from the estimating equation? Per capita income deviations, share of nontraded deviations, and variable exchange rates apparently explain deviations of PL from its norm value of one. African countries might show significantly different behavior, at least for the 1980 ICP data. The equation suggests that forces causing traded-nontraded structures to vary across countries are more important in explaining why the LOP fails.

However, if there are two hypotheses for why the LOP might fail, these two should be tested separately. First, one should test for the failure of the LOP for tradables using exchange rate data, market structure data, and so on. Next, one should test for the failure of LOP for traded-nontraded goods using income, factor endowments, share of nontraded in GDP, and so on. Officer tested the joint hypothesis that the traded-nontrade structure varies across countries and that the traded goods-exchange rate ratio varies across countries. It appears that we can't reject the possibility that the LOP for tradables fails, thus causing PL to deviate from its norm value of one. Separate estimation of the two components of the LOP price--its external and internal versions--is a serious question that clearly opens up a pathway for future research.

3

Real Exchange Rates in Developing Countries: Concepts and Measurement[1]

Sebastian Edwards

1. INTRODUCTION

The exchange rate has been at the center of recent economic debates regarding developing countries. For example, Cline (1989) has argued that the inappropriate exchange rate policies pursued by a number of developing countries in the late 1970s contributed in an important way to the current international debt crisis. Other authors have argued that the maintenance of overvalued exchange rates in Africa for a prolonged period have resulted in the dramatic deterioration observed in that continent's agricultural sector and external position (Gulhati et al., 1985). Still other experts (i.e., Corbo et al., 1986) have postulated that it was the failure to sustain an adequate exchange rate policy that triggered the collapse of the Southern Cone (Argentina, Chile, and Uruguay) experiments with economic reform and free market policies. Moreover, some authors have argued that the economic success of countries like Indonesia, Korea, Thailand and Colombia is to a large extent attributable to the fact that these countries have pursued realistic and appropriate exchange rate policies (Dervis and Petri, 1987).

There is little doubt that during the last 15 years or so, the real exchange rate has claimed a crucial role in the economic literature devoted to economic performance and policies in developing countries. One of the most important of these exchange rate-related problems has to do with defining whether a country's real exchange rate is overvalued or out of line with its long-run equilibrium value. A second important problem refers to how real exchange rates should be measured. Here there are remarkable disagreements, with different people arguing in favor of or against the use of particular indexes.

The purpose of this paper is threefold: first, it discusses the analytical concept of the real exchange rate (RER), placing particular emphasis on providing an operational definition for the equilibrium real exchange rate. Of course, once this concept is defined, we can begin to discuss in a meaningful way what we mean by real exchange rate misalignment, or deviations of the actual RER from its equilibrium value. Second, this paper deals with problems associated with measuring real exchange rates. Several proposals are analyzed and the more serious problems encountered when attempting to compute RERs in the developing countries are discussed. Third, I analyze the actual behavior of developing countries. Here, issues related to the behavior of alternative indexes and to the statistical properties of real exchange rates are emphasized. Additionally, I study the real consequences of increased real exchange rate volatility.[2]

2. EQUILIBRIUM AND DISEQUILIBRIUM REAL EXCHANGE RATES

Exchange rates play a crucial role in determining the external position of a particular country. The long-run external equilibrium position of a country (i.e., current account) will be affected by the <u>real exchange rate</u> as opposed to the <u>nominal</u> exchange rate. In the literature, however, there has been some disagreement regarding the definition of the real exchange rate.[3] In this section some of the alternative definitions offered in the literature are critically reviewed. The concept of "equilibrium" real exchange rate is then introduced, and the difference between equilibrium and disequilibrium real exchange rates is briefly discussed.

2.1. Definition of the Real Exchange Rate

The real exchange rate has been defined in a number of alternative ways in the economic literature. According to earlier views, the real exchange rate was defined as the nominal exchange rate multiplied by the ratio of the foreign to the domestic price level. The main idea was that in an inflationary world changes in the nominal exchange rate would have no clear meaning, and that explicit consideration should be given to changing values in the domestic and foreign currencies, as measured by the respective rates of inflation. In this context a number of writers referred to the real exchange rate as the Purchasing Power Parity (PPP) exchange rate. However, this approach to the real exchange rate is subject to the well-known criticisms and problems of the PPP theory, including those related to the selection of appropriate price indexes and an adequate reference time period.[4]

More recently most authors have defined the real exchange rate in the context of a dependent economy-type model with tradable and nontradable goods. In this setting the real exchange rate has been defined as the (domestic) relative price of tradable to nontradable goods [see, for example, Dornbusch (1974, 1980), Krueger (1978, 1983), Mussa (1979, 1984), and Bruno (1982)]. It should be noted, however, that there is no universally accepted definition of "the" real exchange rate. Indeed, some authors still object to the idea of even considering that an exchange rate--a nominal concept by definition-- could become a real variable [see Maciejewski (1983)], while others continue using the PPP notion of the real exchange rate.

Unless otherwise explicitly stated, in the rest of this study I will use the modern concept of the real exchange rate, defined as the relative price of tradable to nontradable goods. If E is the nominal exchange rate defined as units of domestic currency per unit of foreign currency. P_T^* is the world price of tradables in terms of foreign currency, and P_N is the price of nontradable goods, and no taxes on trade are assumed, the real exchange rate (e) is then defined as:[5]

$$e = \frac{EP_T^*}{P_N}$$

The reason for defining the real exchange rate in this way is that in the context of a tradable and nontradable goods model, the trade account will depend on the (domestic) relative price of tradables to nontradables and not on the PPP definition of the real exchange rate. This follows directly from the fact that the trade account is equal to the excess supply for tradable goods. In fact, assuming that the supply for tradables depends positively on the relative price of those goods (EP_T^*/P_N) and that the demand depends negatively on this relative price and positively on real income, the current account--defined as the excess supply of tradables--will be a positive function of real income and of the relative price of tradables to nontradables or <u>real exchange rate</u>. In this setting, a higher relative price of tradables will result in a higher supply and lower demand for these goods and consequently, assuming that the Marshall-Lerner condition holds, on an improved current account.[6] The real exchange rate defined in this way, then, captures the degree of competitiveness (or profitability) of the tradable goods sector in the domestic country. With other things given, a higher e means a higher degree of competitiveness (and production) of the domestic tradables sector. Williamson (1983a, p. 161) writes: "[I]nternational competitiveness of our goods, ... can

ceteris paribus be identified with the real exchange rate...".
Maciejewski (1983, p. 498), on the other hand, writes: "[S]uch index
values [of the real exchange rate] may provide some broad indication
of the gain or loss in price (cost) competitiveness..." See also Diaz-
Alejandro (1983), Neary and Purvis (1983) and Williamson (1983b).

It is interesting to compare further the tradables-nontradables
relative price definition with the (traditional) PPP definition of the
real exchange rate. The PPP real exchange rate is defined as:

$$e_{PPP} = \frac{EP^*}{P}$$

where P and P* are domestic price indexes. Assuming that these
indexes are geometric weighted averages of tradable and nontradable
prices:

$$P = P_N^{\alpha} P_T^{1-\alpha} \quad ; \quad P^* = P_N^{*\beta} P_T^{*(1-\beta)}$$

and further assuming that the country in question is a small
country and that the law of one price holds for tradable goods
(i.e., $P_T = P_T^* E$), it is possible to find the relation between percentage
changes in e and in the PPP real exchange rate (where, as usual, the
"hat" operator ($^\wedge$) represents a percentage change):

$$\hat{e} = (1/\alpha)\hat{e}_{PPP} + (\beta/\alpha) \, (\hat{P}_T^* - \hat{P}_N^*)$$

From this expression it is possible to see that, in general, changes
in the two definitions of the real exchange rate will differ (i.e., $\hat{e} \neq
e_{PPP}$). Moreover, changes in the two definitions of the real exchange
rate can even go in the opposite direction, depending on the behavior
of foreign relative prices (P_T^* / P_N^*).

2.2. The Equilibrium Real Exchange Rate

From an analytical and policy perspective, a crucial question is
related to the determination of the equilibrium value of the real
exchange rate. Once this equilibrium level is established, it is possible
to determine among other things whether the actual real exchange
rate is misaligned (i.e., overvalued or undervalued).[7] In this section

the literature on the equilibrium real exchange rate is briefly and selectively reviewed.

Robert Mundell (1971) provided an early formal analysis of the determination of the equilibrium real exchange rate. Assuming the case of a small economy that faces given terms of trade, Mundell defines the equilibrium real exchange rate as the relative price of international to domestic goods that simultaneously equilibrates the money market, the domestic goods market and the international goods market. Even though Mundell does not explicitly use the term real exchange rate in this paper, his analysis rigorously describes how the equilibrium relative price of tradables to nontradables is determined.

More recently, Dornbusch (1974, 1980) developed a model of an open dependent economy to analyze the determination of the equilibrium real exchange rate. In its simpler version the model considers a two-goods economy with a tradable and nontradable sector. It is assumed that the production of tradables depends positively on the real exchange rate, while the production of nontradables depends negatively on that relative price. On the other hand, the demand functions for tradables and nontradables are assumed to depend on the real exchange rate and real expenditure. The equilibrium real exchange rate is defined as the relative price of tradables to nontradables at which income equals expenditure, and both the tradables and nontradable goods markets are in equilibrium. Once the equilibrium real exchange rate is defined, Dornbusch investigates the characteristics of disequilibrium in terms of an overvalued or undervalued RER [Dornbusch (1980, pp. 102-103)]. Dornbusch (1980, pp. 103-108) also discusses how, under the assumptions of complete price flexibility and full employment, different disturbances will affect the equilibrium real exchange rate.

A problem with a number of models on the equilibrium real exchange rate, including those of Mundell and Dornbusch, is that they do not allow for a distinction between the effects of temporary and permanent changes in the real exchange rate determinants. This distinction can, in fact, be crucial in some policy discussions. For instance, it is possible to think that while a particular value of the real exchange rate can reflect a short-run equilibrium situation, it may be way out of line with respect to its long-run equilibrium. This possibility recently has been emphasized by a number of authors including Williamson (1983b), Harberger (1983), Edwards (1984), Isard (1983), and Frenkel and Mussa (1984). For example, if there is a <u>temporary</u> transfer from abroad, the real exchange rate that equilibrates the external and internal sectors will appreciate. While this new real exchange rate will be a short-run equilibrium rate--in the sense that it accommodates the transfer--it will be out of line with respect to its equilibrium long-run value (i.e., once the transfer has disappeared).

The distinction between the short-run equilibrium and long-run sustainable equilibrium real exchange rate has been introduced explicitly in some recent analyses of the determination of the equilibrium real exchange rate.[8] In most of these studies the "long-run equilibrium real exchange" rate has been associated with a situation where there is equilibrium in the internal and external sectors <u>and</u> where foreign assets are being accumulated or decumulated at the desired rate. For example, according to Hooper and Morton (1982, p. 43):

> The equilibrium real exchange rate is defined as the rate that equilibrates the current account in the long-run. The long-run equilibrium of "sustainable" current account, in turn, is determined by the rate at which foreign and domestic residents wish to accumulate or decumulate domestic-currency-denominated assets net of foreign currency denominated assets in the <u>long-run</u>.

Williamson (1983b, p. 14) writes:

> [T]he fundamental equilibrium exchange rate is that which is expected to generate a current account surplus or deficit equal to the underlying capital flow over the cycle, given that the country is pursuing international balance as best it can and not restricting trade for balance of payments reasons.

Finally, in their chapter for the <u>Handbook of International Economics</u>, Frenkel and Mussa (1984, p. 64) say that:

> [T]he long-run equilibrium real exchange rate is expected to be consistent with the requirement that on average (in present and future periods), the current account is balanced.

More recently, Neary (1988) has developed an optimizing model of a real economy to analyze the determinants of the equilibrium real exchange rate. An important improvement in Neary's approach over the previous literature is that it explicitly considers that producers and consumers are rational and optimize some objective function. Also, by ignoring all monetary considerations, Neary was able to concentrate on the long-run properties of the model and thus, on the determinants of the equilibrium real exchange rate. A shortcoming of Neary's model, however, is that it is basically static and does not allow distinction between temporary and permanent shocks or between anticipated and unanticipated disturbances.

2.3. An Intertemporal Model of Equilibrium Real Exchange Rates

In this subsection a minimal fully optimizing model of equilibrium real exchange rates is presented. The model is partially based on Edwards (1989a,b). The equilibrium real exchange rate (ERER) is defined as that relative price of tradables to nontradables that, for given sustainable (equilibrium) values of other relevant variables such as taxes, international prices and technology, results in the simultaneous attainment of internal and external equilibrium. Internal equilibrium means that the nontradable goods market clears in the current period and is expected to be in equilibrium in future periods. Implicit in this definition of the equilibrium RER is the idea that this equilibrium takes place with unemployment at the "natural" level. External equilibrium, on the other hand, is attained when the intertemporal budget constraint that states that the discounted sum of a country's current account must equal zero is satisfied. In other words, external equilibrium means that current account balances (current and future) are compatible with long-run sustainable capital flows.

A number of important implications follow from this definition of equilibrium real exchange rate. First, the ERER is not an immutable number. When there are changes in any of the other variables that affect the country's internal and external equilibria, there will also be changes in the equilibrium real exchange rate. For example, the RER "required" to attain equilibrium will not be the same with a very low world price of the country's main export as it would with a very high price of that good. In a sense then, the ERER is itself a function of a number of variables including import tariffs, export taxes, real interest rates, capital controls and so on. These immediate determinants of the ERER are the real exchange rate "fundamentals." Second, the ERER will be affected not only by current "fundamentals" but also by the expected future evolution of these variables. To the extent that there are possibilities for intertemporal substitution of consumption via foreign borrowing and lending and of intertemporal substitution in production via investment, expected future events--such as an expected future change in the international terms of trade, for example--will have an effect on the current value of the ERER. In particular, the behavior of the equilibrium real exchange rate will depend on whether changes in fundamentals are perceived as being permanent or temporary. If there is perfect international borrowing, a temporary disturbance to, say, the terms of trade, will affect the complete future path of equilibrium RERs. However, if there is rationing in the international credit market, intertemporal substitution through consumption will be cut and temporary disturbances will tend to affect the ERER in the short run only. In this case a distinction between short-run and long-run equilibrium real exchange rates becomes useful.

Although this framework is very general and it can accommodate many goods and factors, it is useful to think of this small economy as being comprised of a large number of profit-maximizing firms, that produce three goods--exportables (X), importables (M) and nontradables (N)--using constant returns to scale technology under perfect competition. It is assumed that there are more factors than tradable goods so factor price equalization does not hold. One way to think about this is by assuming that each sector uses capital, labor and natural resources.

There are two periods only--the present (period 1) and the future (period 2)--and there is perfect foresight. Residents of this small country can borrow or lend internationally. There are, however, taxes on foreign borrowing; the domestic (real) interest rate exceeds the world interest rate. The intertemporal constraint states that at the end of period 2 the country has paid its debts. The importation of M is subject to specific import tariffs in both periods 1 and 2. In this model the current account is equal to savings minus investment in each period. Consumers maximize intertemporal utility and consume all three goods.

There is a government that consumes both tradables and nontradables. Government expenditure is financed through nondistortionary taxes, proceeds from import tariffs, proceeds from the taxation of foreign borrowing by the private sector, and borrowing from abroad. As in the case of the private sector, the government is subject to an intertemporal constraint: the discounted value of government expenditure (including foreign debt service) has to equal the discounted value of income from taxation.

In addition to the private sector and government budget constraints, internal equilibrium requires that the nontradable market clear in each period. That is, the quantity supplied of nontradables has to equal the sum of the private and public sectors' demands for these goods. The model is completely real; there is no money or other nominal assets.

The general model is given by equations (1) through (9), where the (world) price of exportables has been taken as the numeraire:

$$R(1,p,q,V,K) + \delta \tilde{R}(1,\tilde{p},\tilde{q},\tilde{V},K+I) - I(\delta) - T - \delta \tilde{T} = \tag{1}$$

$$E(\pi(1,p,q),\delta \tilde{\pi}(1,\tilde{p},\tilde{q}),W),$$

$$G_X + p{*}G_M + qG_N + \delta{*}(\tilde{G}_X + \tilde{p}{*}\tilde{G}_M + \tilde{q}\tilde{G}_N) - \tau(E_p - R_p) + \tag{2}$$

$$\delta{*}\tilde{\tau}(E_{\tilde{p}} - R_{\tilde{p}}) + b(NCA) + T + \delta{*}\tilde{T},$$

$$R_q - E_q + G_N, \tag{3}$$

$$\tilde{R}_{\tilde{q}} - E_{\tilde{q}} + \tilde{G}_N, \tag{4}$$

$$p - p{*} + \tau, \tag{5}$$

$$\tilde{p} - \tilde{p}{*} + \tilde{\tau}, \tag{6}$$

$$\delta \tilde{R}_K - 1, \tag{7}$$

$$P_T^{*} - \tau P_M^{*} + (1 - \tau)P_X^{*}; \quad \tilde{P}_T^{*} - \tau \tilde{P}_M^{*} + (1-\tau)\tilde{P}_X^{*}; \quad (P_X^{*} - \tilde{P}_X^{*} - 1) \tag{8}$$

$$RER - (P_T^{*}/P_N); \quad \tilde{R}ER - (\tilde{P}_T^{*}/\tilde{P}_N) \tag{9}$$

Table 3.1 contains the notation used.

Equation (1) is the intertemporal budget constraint for the private sector and states that present value of income valued at domestic prices must equal present value of private expenditure. Given the assumption of a tax on foreign borrowing, the discount factor used in (1) is the domestic factor δ, which is smaller than the world discount factor δ^{*}.

Equation (2) is the government intertemporal budget constraint. It states that the discounted value of government expenditure has to equal the present value of government income from taxation. NCA, which is equal to $(\tilde{R}-\tilde{\pi}E_{\tilde{q}})$ in (2), is the private sector current account surplus in period 2, and b(NCA) is the discounted value of taxes on foreign borrowing paid by the private sector. Notice that the use of the world discount factor δ^{*} in (2) reflects the assumption that in this model the government is not subject to the tax on foreign borrowing.

Equations (3) and (4) are the equilibrium conditions for the nontradables market in periods 1 and 2; in each of these periods the quantity supplied of N (R_q and $\tilde{R}_{\tilde{p}}$) has to equal the sum of the

TABLE 3.1. Notation Used in Model of Equilibrium Real
Exchange Rates

$R(\ )$; $\tilde{R}(\ )$	Revenue functions in periods 1 and 2. Their partial derivatives with respect to each price are equal to the supply functions.
p; $\tilde{p}$	Domestic relative price of importables in periods 1 and 2.
q; $\tilde{q}$	Relative price of nontradables in periods 1 and 2.
V; $\tilde{V}$	Vector of factors of production, excluding capital.
K	Capital stock in period 1.
$I(\ )$	Investment in period 1.
$\delta*$	World discount factor, equal to $(1+r*)^{-1}$, where $r*$ is world real interest rate in terms of exportables.
δ	Domestic discount factor, equal to $(1+r)^{-1}$. Since there is a tax on foreign borrowing, $\delta < \delta*$.
$b = (\delta*-\delta)$	Discounted value of tax payments per unit borrowed from abroad.
$p*$; $\tilde{p}*$	World relative price of imports in periods 1 and 2.
r; $\tilde{\gamma}$	Import tariffs in periods 1 and 2.
T; $(\tilde{T})$	Lump sum tax in periods 1 and 2.
G_X, G_M, G_N; $\tilde{G}_X, \tilde{G}_M, \tilde{G}_N$	Quantities of goods X, M and N consumed by government in periods 1 and 2.
$E(\ )$	Intertemporal expenditure function.

Table 3.1 (continued)

$\pi(1,p,q)$; $\tilde{\pi}(\ )$	Exact price indexes for periods 1 and 2, which under assumptions of homotheticity and separability correspond to unit expenditure functions.
W	Total welfare.
NCA	Noninterest current account of the private sector in period 2.
P_M^*, P_X^*; $\tilde{P}_M^*, \tilde{P}_X^*$	Nominal world prices of M and X in periods 1 and 2. Notice that we assume that $P_X^* = \tilde{P}_X^* = 1$.
P_N; P_N	Nominal price of nontradables in periods 1 and 2.
P_T^* ; $\tilde{P}_T^*$	World prices of tradables, computed as an index of the prices of X and M, in periods 1 and 2.
RER;RÉR	Definition of the real exchange rate in periods 1 and 2.Indicates the value of a variable in period 2.

---

quantity demanded by the private sector (E_q and $E_{\bar{q}}$) and by the government. Given the assumptions about preferences (separability and homotheticity), the demand for N by the private sector in period 1 can be written as:

$$E_q = \pi_q E_{\tau} \qquad (10)$$

Equations (5) and (6) specify the relation between domestic prices of importables, world prices of imports, and tariffs. Equation (7) describes investment decisions, and states that profit-maximizing firms will add to the capital stock until Tobin's "q" equals 1. This expression assumes that the stock of capital is made up of the numeraire good.

In this model I can distinguish between the "exportables real exchange rate" (1/q) and the "importables real exchange rate" (p/q). Since the relative price of X and M can change, one cannot really talk about a tradable goods composite. It is still possible, however, to <u>compute</u> how an index of tradables prices evolves through time. Equation (8) is the definition of the price index for tradables, where τ and $(1-\tau)$ are the weights of importables and exportables. Equation (9) defines the real exchange rate index as the domestic relative price of tradables to nontradables. Equations (1) through (9) fully describe the inter- and intratemporal (external or internal) equilibria in this economy.

In this model there is an equilibrium path for the RER. The vector of equilibrium RERs, **RER** = (RER, RER), is composed of those RERs that satisfy equations (1) through (9) for given values of the other fundamental variables. Notice that since no rigidities, externalities, or market failures are assumed, the equilibrium real exchange rates imply the existence of "full" employment [see, however, Edwards (1989a)].

From inspection of equations (1)-(9), it is apparent that exogenous shocks in, say, the international terms of trade will affect the vector of equilibrium relative prices and RERs through two interrelated channels. The first one is related to the <u>intratemporal</u> effects on resource allocation and consumption and production decisions. For example, as a result of a temporary worsening of the terms of trade, there will be a tendency to produce more and consume less of M in that period. This, plus the income effect resulting from the worsening of the terms of trade will generate an incipient disequilibrium in the nontradables market that must be resolved by a change in relative prices and in the equilibrium RER. In fact, if we assume that there is an absence of foreign borrowing, these intratemporal effects will be the only relevant ones. However, with capital mobility and investment

as in the current model, there is an additional <u>inter</u>temporal channel
through which changes in exogenous variables will affect the vector of
equilibrium RERs. For example, in the case of a worsening of the
terms of trade, the consumption discount factor $\bar{\pi}\delta/\pi$ will be affected,
altering the intertemporal allocation of consumption. Also, in that
case the investment equilibrium condition (7) will be altered, affecting
future output.

Naturally, without specifying the functional forms of the
expenditure, revenue, and other functions in (1)-(9), it is not possible
to write the vector of equilibrium relative prices of nontradables or
the equilibrium real exchange rates in an explicit form. It is possible,
however, to write them implicitly as functions of all the sustainable
levels of all exogenous variables (contemporaneous and anticipated)
in the system:

$$\text{RER} = h(p*, \; \tilde{p}*, \; s, \; \tilde{c}, \; d, \; d*, \; V, \; T, \; \tilde{T}, \; G_\chi, \; \tilde{G}_\chi, \; \ldots) \qquad (11)$$

$$\text{RER} = h(p*, \; \tilde{p}*, \; s, \; \tilde{c}, \; d, \; d*, \; V, \; T, \; \tilde{T}, \; G_\chi, \; \tilde{G}_\chi, \; \ldots) \qquad (12)$$

A crucial question is related to the way in which the equilibrium
vectors of relative prices and RERs will change in response to
different types of disturbances. That is, we are interested in the (most
plausible) signs of the partial derivatives of RER and RER with
respect to their determinants. The actual discussion of these effects
is beyond the scope of this paper, and can be found in Edwards
(1989a,b,c). The main conclusions from the manipulation of the
model can be summarized as follows:

(1) With low initial tariffs the imposition of import tariffs (either
temporarily or permanently) will usually generate an <u>equilibrium real
appreciation</u> in the current and future periods. A sufficient condition
is that we have (net) substitutability in demand among all three goods
X, M and N. If initial tariffs are high, for this result to hold, we need,
in addition, income effects to be dominated by substitution effects. If,
however, there is complementarity in consumption, it is possible that
the imposition of import tariffs will generate a real equilibrium
depreciation.

(2) If the income effect associated with a terms of trade deterioration
dominates the substitution effect, a worsening in the terms of trade
will result in an <u>equilibrium real depreciation</u>.

(3) Generally speaking, it is not possible to know how the effect of
import tariffs and terms of trade shocks on the ERER will be
distributed through time.

(4) It is crucially important to distinguish between permanent and temporary shocks when analyzing the reaction of the equilibrium real exchange rate.

(5) A relaxation of exchange controls will always result in an equilibrium real appreciation in period 1. Moreover, in that period we will observe simultaneously a real appreciation and an increase in borrowing from abroad.

(6) A transfer from the rest of the world--or an exogenously generated capital inflow for that matter--will always result in an equilibrium real appreciation.

(7) The effect of an increase in government consumption on the equilibrium RERs will depend on the composition of this new consumption. If it falls fully on nontradables, there is a strong presumption that the RER will experience an equilibrium real appreciation. If it falls fully on tradables, there will be an equilibrium real depreciation.

2.4. Real Exchange Rate Misalignment

Even though, as suggested by the model presented above, long-run equilibrium real exchange rates are a function of real variables only, actual real exchange rates respond both to real and monetary variables. The existence of an equilibrium real exchange rate does not mean that the actual real exchange rate has to be permanently equal to this equilibrium value. In fact, the actual RER will normally exhibit departures from its long-run equilibrium; short-run and even medium-run deviations of the actual from the equilibrium RER that typically are not very large and that stem from short-term frictions and adjustment costs can be quite common. However, there are other types of deviations that can become persistent through time, generating major and sustained differentials between actual and equilibrium real exchange rates or real exchange rate misalignments.

To construct a model of real exchange rate misalignment, it is necessary to abandon the frictionless "real" world of the previous subsection; it is necessary to introduce monetary and financial sectors as well as rigidities that impede instantaneous adjustments. Although the construction of such a model is well beyond the scope of this paper, in the rest of this section some of the most important characteristics of misalignment situations will be discussed.

A fundamental principle of open economy macroeconomics is that in order to have a sustainable macroeconomic equilibrium, monetary and fiscal policies must be consistent with the chosen nominal exchange rate regime. This means that the selection of an exchange rate system imposes certain limitations on the extent of macropolicies. If this consistency is violated, severe disequilibrium situations, which are usually reflected on real exchange rate misalignment, will take place.

Perhaps the case of a "high" fiscal deficit under fixed nominal exchange rates is the most clear example of macro and exchange rate inconsistencies. In most developing countries fiscal imbalances are partially or wholly financed by money creation. The inflation required to finance a fiscal deficit equal to a fraction δ of GDP can be calculated as:

$$\pi = \delta/\lambda \qquad\qquad (13)$$

where π is the rate of inflation required to finance the government deficit, and λ is the ratio of high-powered money to GDP. If the required rate of inflation is "too high," it will possibly result in the price of nontradables (P_N) growing faster than the international price of tradables (P_T^*) and in real appreciation. This type of "inconsistent" fiscal policy will result in domestic credit creation above money demand growth. This, in turn will be translated into an excess demand for tradable goods, nontradable goods, and financial assets. While the excess demand for tradables will be reflected in a higher trade deficit (or lower surplus), in a loss of international reserves, and in an increase in (net) foreign borrowing above its long-run sustainable level, the excess demand for nontradables will be translated into higher prices for those goods and consequently into a real exchange rate appreciation. If there are no changes in the fundamental real determinants of the equilibrium RER, this real appreciation induced by the expansive domestic credit policy will represent a departure of the actual RER from its equilibrium value or real exchange rate misalignment. Naturally, since this policy is unsustainable, something will have to give. Either the inconsistent macropolicies will have to be reverted or at some time the central bank will "run out" of reserves and a balance of payments crisis will ensue.

The consistency between monetary and exchange rate policies is needed not only under fixed rates but also under most types of predetermined and managed nominal exchange rates such as an active crawling peg. Perhaps the situation in Argentina in the late 1970s is the most notorious recent case of inconsistent fiscal and crawling nominal exchange rate policies. During that period the Argentinian government implemented the by-now-famous preannounced rate of devaluation of "tablita" as a means to reduce inflation. However, the preannounced rate of crawl clearly was inconsistent with the inflation tax required to finance the fiscal deficit (Calvo 1986). This inconsistency generated not only a real appreciation but also a substantial speculative activity where the public basically bet on when the "tablita" would be abandoned.

Nonunified (or multiple) nominal exchange rates traditionally have had some appeal for the developing countries and recently have become fairly common. Under this type of system different international transactions are subject to differential nominal exchange rates, giving rise to the possibility of the existence of more than one real exchange rate. Under nonunified exchange rates, the relation between macroeconomic policies and the rest of the economy will depend on the nature of the multiple rates system. if, for example, the multiple rates regime consists of two (or more) predetermined (i.e., fixed) nominal rates, the system will work in almost the same way as under unified predetermined nominal rates. This is because multiple fixed nominal exchange rates are perfectly equivalent to a unified rate system with taxes on certain external transactions. In this case, as with unified predetermined rates, inconsistent macroeconomic policies will result in loss of international reserves, a rate of domestic inflation that will exceed world inflation, and in real exchange rate overvaluation. This situation, of course, will be unsustainable in the long run and the authorities will have to introduce corrective macropolicies.

A different kind of nonunified nominal exchange rate consists of a fixed official rate for current account transactions and an (official) freely fluctuating rate for capital account transactions. The main purpose of this system is to delink the real side of the economy from the effects of supposedly highly unstable capital movements. In this dual exchange rate system, portfolio decisions are highly influenced by the differential between the free and fixed rates or exchange rate premium. The private sector decisions on what proportion of wealth to hold in the form of foreign currency-denominated assets is directly influenced by the expected rate of devaluation of the free rate.

Under a dual exchange rate regime, even if no current account transactions slip into the free rate, changes in the fluctuating nominal rate will influence the real exchange rate. Consider, for example, the case of an increase of domestic credit at a rate that exceeds the increase in the demand for domestic money. As before, this will provoke an excess demand for goods and financial assets. As a result of this policy, there will be a decline in the stock of international reserves, an increase in the price of nontradable goods and consequently, a real appreciation. In addition, there will be an increase in the demand for foreign assets that will result in a nominal devaluation of the free rate and in changes in the domestic interest rate. The devaluation of the free rate will, in turn, have secondary effects on the official <u>real</u> exchange rate via a wealth effect. The bottom line, however, is that in this case inconsistent macropolicies also eventually will be unsustainable, as international reserves are drained. By partially delinking the current from the capital account, all the dual rates system can hope to do is delay the eventual crisis. A system that is particularly relevant for the developing countries consists of the coexistence of a fixed rate for commercial transactions with a floating <u>parallel</u> (either black or grey) market rate governing the financial transactions.[9]

3. MEASURING REAL EXCHANGE RATES

From an empirical point of view, the first question that should be addressed is: How should the real exchange rate be measured? From equation (9), which defines the real exchange rate as the relative price of tradables and nontradables, it is apparent that the main measurement problems are those related to the selection of the real-world counterparts of P_T^* and P_N. In reality, it is extremely difficult, if not impossible, to define which goods are actually tradables and which are nontradables. A second measurement problem is related to the definition of E. Should the nominal exchange rate with respect to the U.S. dollar be considered? Or is the exchange rate with respect to the DM the most appropriate? Or should an average of both rates be used? These and other problems related to the measurement of the real exchange rate will be discussed in this section. The analysis will be restricted to the actual measurement of the RER without entering into the important and difficult question of the empirical definition of the equilibrium level of the real exchange rate. The analysis presented in this section will first discuss briefly the arguments traditionally given favoring alternative measures of the real exchange rate. The discussion will be quite general and will provide a broad cover of the literature. That is, the presentation will also deal, even though briefly, with the PPP real exchange rate. Section 4, on the other hand, deals with the actual behavior of different RER indexes in the developing countries.

As expressed in equation (9), RER is defined as the relative price of tradable to nontradable goods. Ideally, one would want to have data on tradables and nontradables. In almost every country, however, these are not available. For this reason, some proxy for the analytical concept of the RER should be found. In some respects, the selection of the appropriate proxies for P_N and P_T^* resembles the definition of the adequate price levels in the discussions of the Purchasing Power Parity theory [see, for example, Keynes (1924), Viner (1937), and Officer's (1976) review]. Indeed, most of the discussion on the appropriate measurement of the real exchange rate has been closely related to the PPP literature.

Basically, four alternative price indexes traditionally have been suggested as possible candidates for the construction of the real exchange rate index. However, as we will see, most of these propositions relate to the traditional PPP definition, and are not entirely appropriate as proxies for the relative price of tradables to nontradables. The following price indexes actually have been suggested: (1) the Consumer price Indexes at home and abroad (CPI); (2) the Wholesale Price Indexes (WPI); (3) the GDP deflators (GD); (4) and wage rate indexes (WR).[10] Also, some authors have suggested using specific components of the CPI and WPI as proxies for the prices of tradables and nontradables. In practice, however, this procedure has the same types of problems as those arising from the use of more standard price indexes. The relative merits of these indexes are also somewhat related to the old PPP discussion.

Of course, none of these indexes are perfect and all of them present some advantages and disadvantages. The relevant question, then, is which index, or indexes, are preferable for analyzing changes in the real exchange rate and the degree of competitiveness. In the rest of this subsection the discussion will be restricted to the merits of the alternative price indexes. Below, the question of bilateral versus multilateral real exchange rates will be tackled in detail.

Within the context of the PPP real exchange rate, the most commonly used index of the real exchange rate in empirical and policy discussions is that constructed using CPIs as the relevant price indexes [see deVries (1968)]. It has been argued that this indicator will provide a comprehensive measure of changes in competitiveness, since the CPIs include a broad group of goods, including services [see Genberg (1978)]. Another advantage of this index is that almost every country periodically (i.e., monthly) publishes fairly reliable data on CPI behavior. However, an obvious problem with this measure is that since the CPI includes a large number of nontraded goods, it will tend to provide a biased measure of the changes in the degree of competitiveness of the tradable goods sector [see Frenkel (1978), Officer (1982)].

Some authors have suggested that this problem would be solved if WPI indexes, which contain mainly tradable goods are used in the computation of the real exchange rate. This measure, however, has also been subject to criticism. It has been argued, for example, that since these indexes contain highly homogeneous tradable goods, whose prices tend to be equated across countries when expressed in a common currency, the real exchange rate computed using WPIs will vary very little without really measuring actual changes in the degree of competitiveness [see Keynes (1930), Officer (1982)].[11] Also, the use of WPI (as well as other) indexes is subject to the problem arising from the use of different weights across countries.

The main merit of the GDP deflator as a candidate for the construction of the RER is that it is a genuine price index of aggregate production, while both the CPI and the WPI are indexes of consumption prices. It has been thought then, that a real exchange rate index computed using GDP deflators will provide a good indicator of changes in the degree of competitiveness in production [see Officer (1976, 1982); Barro (1983)]. On the other hand, a crucial drawback of the GDP deflator is that for most developing countries it is only available on a yearly basis and that, as in the case of the CPI, it has a large component of nontradable goods [see Harberger (1981)].

Many authors, including the IMF staff [Artus (1978), Artus and Knight (1984)], prefer to compute the real exchange rate as a ratio of unit labor costs [see also Houthakker (1962, 1963)], the reason being that this index is, in some sense, a direct measure of relative competitiveness across countries [see Maciejewski (1983)]. It has also been argued that relative labor costs are more stable than relative goods prices [Artus (1978), Officer (1982)]. As in the case of the other indexes, there are a number of analytical problems related to the use of this type of measure for the real exchange rate. First, an indicator based on wage rates behavior will be highly sensitive to cyclical productivity changes. For this reason the IMF has constructed the so-called <u>normalized unit labor costs</u> indexes which correct the competitiveness measure by these productivity changes [see Maciejewski (1983), and <u>International Financial Statistics</u> (April 1984, p. 63)]. Unfortunately, however, because of data availability limitations, the IMF computes these normalized unit labor costs only for the <u>OECD</u> countries. A second shortcoming of the wage rate-based measure of the real exchange rate is that it takes into account only one factor of production. To the extent that the capital/labor ratio differs across countries, this will introduce a bias into the index. Finally, the poor quality and limited availability of wage rate data for developing countries is also a serious drawback in using this indicator.

Recently some authors have argued that the best way to construct a real exchange rate index is to use some component of the more traditional price indexes to construct proxies for the domestic price of tradables and nontradables. For example, Kravis and Lipsey (1983) have suggested using (for most countries) the GDP deflator for services and government expenditures to construct a proxy for nontradables and the deflators of the rest of the sectors to construct a proxy for tradables. Even though this sounds like a sensible proposition, it has two important drawbacks. First, the existing disaggregation at the national account level in most countries is too broad to allow for really meaningful comparisons across sectors. Second, and more important, with very few exceptions, national account data are only available on a yearly basis and with a substantial delay. This, unfortunately, defeats the whole idea of having a reliable and fast index of external competitiveness. At this level, a more practical proposition is to construct the real exchange rate using components of the consumer and wholesale price indexes to build the proxies for tradables and nontradables prices. These indexes are available fairly quickly for almost every country on a monthly basis. A problem with this proposition, however, is how to select of which components to include as part of what index. Another problem, of course, is related to selection of the weight to attach to each component in construction of the proxies. Even though these are tricky problems, they are not insurmountable. Their solution basically will require good judgment.

Thus, from a practical point of view and for most purposes, it is advisable to stick to real exchange rate indexes constructed with the traditional price indexes. There are two main advantages to this. First, the cost involved in building these series is relatively low; second, in this way cross-country comparisons can be made more easily. In the rest of this section, the discussion will be restricted to the behavior of real exchange rate indexes constructed using CPIs, WPIs, GDP deflators and wage indexes. A growing number of authors recently have proposed that an adequate proxy for the relative price of tradables to nontradables can be constructed if the foreign WPI is used in the numerator and the domestic CPI is used in the denominator. Later in this section a more detailed discussion on the merits and demerits of this particular index will be provided.

3.1. The Real Exchange Rate in a World of Floating: Effective Real Exchange Rates vs. Bilateral Real Exchange Rates

The preceding discussion referred to <u>bilateral</u> rates between the domestic currency and, say, the U.S. dollar. However, in a world in which the main currencies are floating, there are many different bilateral rates and there is no reason to prefer one rate over another. For this reason, indexes of real exchange rates have been constructed that take into account the behavior of all the relevant bilateral rates. These exchange rate indexes have been called <u>real effective exchange rates</u> or <u>real basket exchange rates</u>.

The behavior of the effective exchange rate can be, at least in theory, very different from the behavior of any bilateral exchange rate. In order to illustrate this point, we will concentrate on effective vs. bilateral nominal exchange rates. The analysis follows easily for the case of real exchange rates. Assume that a country trades with k countries. Then the effective nominal exchange rate B_t is defined as:

$$B_t = \sum_{i=1}^{k} \alpha_i E_{cit} \tag{14}$$

where α_i is the appropriate weight for country i, and E_{ci} is an index of the bilateral nominal exchange rate between the home country's currency and country i's currency in period t.[12] By triangular arbitrage:

$$E_{ci} = E_{c1} E_{1i} \quad i = 1, 2, \ldots, k$$

where E_{c1} is, for example, the bilateral nominal exchange rate between the home country and the U.S. dollar, and E_{1i} is the rate between the U.S. dollar and country i's currency (i.e., the U.S. dollar/Yen rate).

The rate of change of the nominal effective exchange rate B_t can be written as (where $\hat{X} = dX/dt \, 1/X$):

$$\hat{B} = \hat{E}_{c1} + \left[\sum_{i=2}^{k} \left(\frac{\alpha_i E_{1i}}{A} \right) \hat{E}_{1i} \right] \tag{15}$$

where

$$A = \alpha_1 + \sum_{j=2}^{k} \alpha_j E_{1j}$$

Equation (15) indicates that in a world of floating rates the rate of change of the <u>effective nominal rate</u> $\hat{\beta}$ will differ from the change in the bilateral rate with respect to the reference country $\hat{E}_{c1}$ by the term in square brackets. In particular, if the U.S. dollar--the currency in terms of which the bilateral rate is defined--is appreciating in the world market (i.e.,

$$\sum_{i=2}^{k} (\alpha_i E_{1i}/A) \, \hat{E}_{1i} < 0),$$

of the rate of nominal depreciation the effective nominal rate will be smaller than the rate of nominal depreciation of the bilateral rate ($\hat{B} < \hat{E}_{c1}$). Of course, the contrary would be the case if the dollar depreciated relative to the other currencies, as has been the case since 1985.

4. REAL EXCHANGE RATE BEHAVIOR IN SELECTED DEVELOPING COUNTRIES

The purpose of this section is to investigate how different indexes of RERs have behaved in a large number of developing countries. In particular, the analysis compares the behavior of bilateral, effective, official and parallel market exchange rates. The discussion concentrates on determining whether RERs have exhibited trends and on how volatile RERs have been. In addition, the analysis inquires into whether it really makes a difference which price indexes are used to construct RER indexes.

4.1. Official Nominal Exchange Rates and RER Behavior in 33 Developing Countries

Effective Real Exchange Rates and Bilateral Exchange Rates

In the construction of the effective indexes of real exchange rates, the following equation was used:

$$MRER_{jt} = \frac{\overset{k}{\underset{i=1}{R}} a_i E_{it} P^*}{P_{jt}}$$

where $MRER_{jt}$ is the index of the multilateral or effective real rate in period t for country j; E_{it} is an index of the nominal rate between country i and country j in period t; $i = 1,...,k$ refers to the k partner countries used in construction of the MRER index; α_i is the weight corresponding to partner i in the computation of $MRER_{jt}$; P^*_{it} is the price index of the i partner in period t; and P_{jt} is the price index of the home country in period t. An increase in the value of this index of MRER reflects a real depreciation, whereas a decline implies a real appreciation of the domestic currency.

Two indexes of multilateral real exchange rates were constructed and their behavior compared. The first index--a proxy for the relative price of tradables to nontradables--used the partner countries' WPIs as the P_{it}^*'s and the home country CPI as P_{jt}. For notation purposes this index was called MRER1. The second index--which is related to the more traditional PPP measure of the real exchange rate--used consumer price indexes for both partner countries and the home country. This index was called MRER2.

In constructing both indexes, the following procedure was followed: (1) The weights (α's) were trade weights constructed using data from the International Monetary Fund's Directions of Trade. (2) For each country the ten largest trade partners in 1975 were used for construction of the real exchange rate indexes. (3) In all cases the nominal exchange rate indexes (E_{ij}) were constructed from data on official nominal exchange rates obtained from the International Financial Statistics (IFS). In those cases in which there were multiple official exchange rates, the "most common" rate as listed by IFS was used. This means that these indexes are capturing some of the distortions introduced by the existence of multiple rates. What they do not capture, however, is the role of nonofficial black or parallel markets for foreign exchange [see, however, Edwards (1989a)].

Two indexes of bilateral real exchange rates with respect to the United States were also constructed using data on official nominal rates. These indexes were defined as:

$$BRER1 = \frac{E\ WPI^{US}}{CPI}$$

and

$$BRER2 = \frac{E\ CPI^{US}}{CPI}$$

where E is the bilateral (official) nominal exchange rate with respect to the U.S. dollar; WPI^{US} and CPI^{US} are the wholesale and consumer price indexes; and CPI is the domestic country consumer price index. BRER1, then, is the bilateral counterpart of MRER1. On the other hand, BRER2 uses both the domestic country and U.S. CPIs and historically has been the most popular RER index in policy analysis.

Figures 3.1 through 3.8 show the evolution of two real exchange rate indexes, the multilateral MRER1 index and the bilateral BRER1 index, for 33 developing countries. As may be seen, in most cases both indexes tended to move roughly in the same direction throughout most of the period, and in particular between 1960 and 1971. After the collapse of the Bretton Woods system, in many of the countries depicted in these diagrams the multilateral and bilateral indexes started to exhibit some differences in behavior.

To formally compare the behavior of the four alternative indexes of the real exchange rate constructed using official data, coefficients of correlation between the multilateral and the bilateral real exchange rate indexes were computed using quarterly data for the period from the first quarter of 1965 to the second quarter of 1985. The following regularities emerged from this analysis. First, in most countries the two alternative definitions of the bilateral real exchange rate index moved closely during this period. In 27 of the 33 countries considered, the coefficient of correlation between log(BRER1) and log (BRER2) was above 0.9, and in all cases it exceeded 0.8. Second, the two indexes of trade-weighted multilateral RER also moved closely. In 30 of the 33 countries, the coefficient of correlation between the logs of MRER1 and MRER2 exceeded 0.9. Third, the behavior of the bilateral and multilateral RER indexes has been quite different in many of these countries. In 16 cases the coefficient of correlation between log MRER and log BRER was below 0.5 and in two countries it even was negative. These findings indicate that for most countries, and within a particular type of index--bilateral or multilateral--the selection of the price indexes used in the construction of the RER measure is not a major practical problem. The results also show that the bilateral and multilateral real exchange rate indexes move in different, and even opposite, directions. This means that when evaluating policy-related situations, it is necessary to use or construct a broad multilateral index of real exchange rate. Failure to do this can result in misleading and incorrect inferences regarding the evolution of a country's degree of competitiveness.

Trends and Variability

The real exchange rate indexes depicted in Figures 3.1 through 3.8 have two important characteristics. First they show that in most countries the real exchange rate has been fairly variable. Second, in spite of the observed variability, in several of these countries it

Figure 3.1

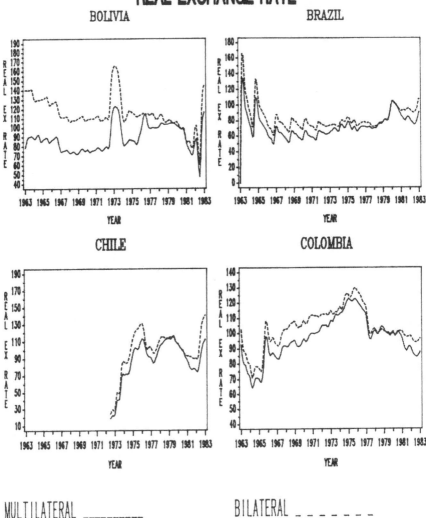

MULTILATERAL AND BILATERAL
REAL EXCHANGE RATE

MULTILATERAL _____   BILATERAL _ _ _ _ _ _ _
SOURCE: CONSTRUCTED FROM RAW DATA OBTAINED FROM THE I.F.S.

81

Figure 3.2

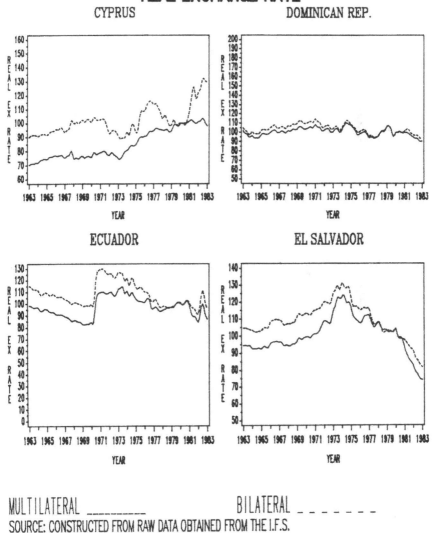

MULTILATERAL AND BILATERAL
REAL EXCHANGE RATE

MULTILATERAL _____   BILATERAL _ _ _ _ _ _ _
SOURCE: CONSTRUCTED FROM RAW DATA OBTAINED FROM THE I.F.S.

Figure 3.3

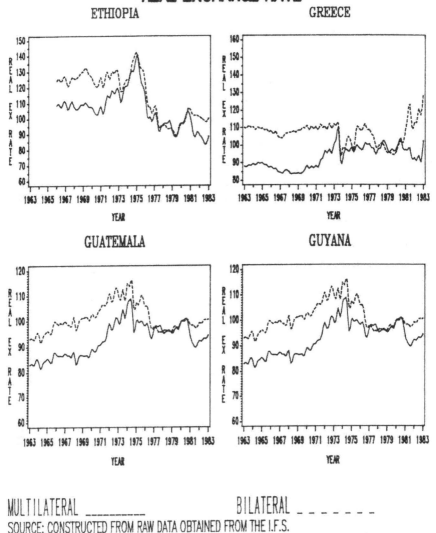

MULTILATERAL AND BILATERAL
REAL EXCHANGE RATE

MULTILATERAL _____   BILATERAL _ _ _ _ _ _

SOURCE: CONSTRUCTED FROM RAW DATA OBTAINED FROM THE I.F.S.

Figure 3.4

MULTILATERAL AND BILATERAL
REAL EXCHANGE RATE

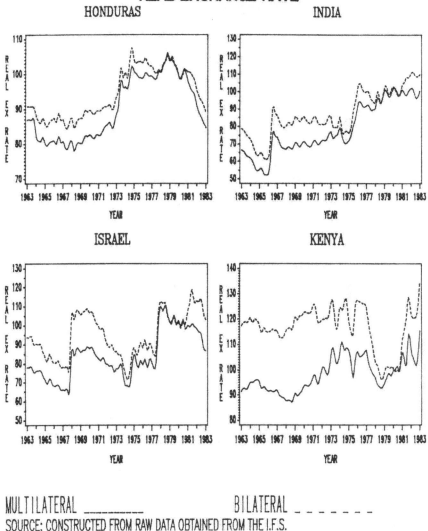

MULTILATERAL _____   BILATERAL _ _ _ _ _ _ _
SOURCE: CONSTRUCTED FROM RAW DATA OBTAINED FROM THE I.F.S.

Figure 3.5

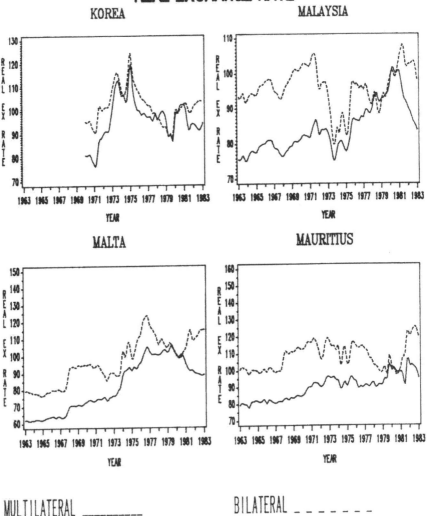

MULTILATERAL AND BILATERAL
REAL EXCHANGE RATE

MULTILATERAL _____ BILATERAL _ _ _ _ _ _
SOURCE: CONSTRUCTED FROM RAW DATA OBTAINED FROM THE I.F.S.

Figure 3.6

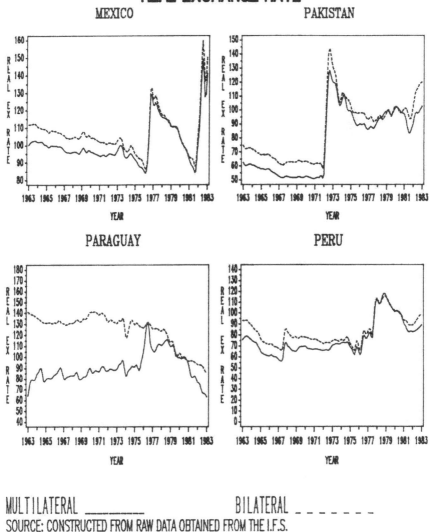

MULTILATERAL AND BILATERAL
REAL EXCHANGE RATE

MULTILATERAL _____   BILATERAL _ _ _ _ _ _
SOURCE: CONSTRUCTED FROM RAW DATA OBTAINED FROM THE I.F.S.

Figure 3.7

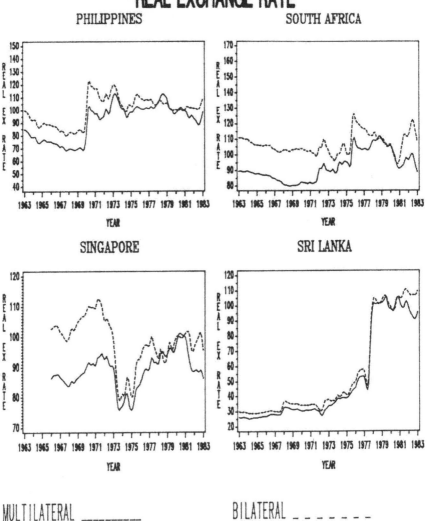

MULTILATERAL AND BILATERAL
REAL EXCHANGE RATE

MULTILATERAL _____   BILATERAL _ _ _ _ _ _

SOURCE: CONSTRUCTED FROM RAW DATA OBTAINED FROM THE I.F.S.

Figure 3.8

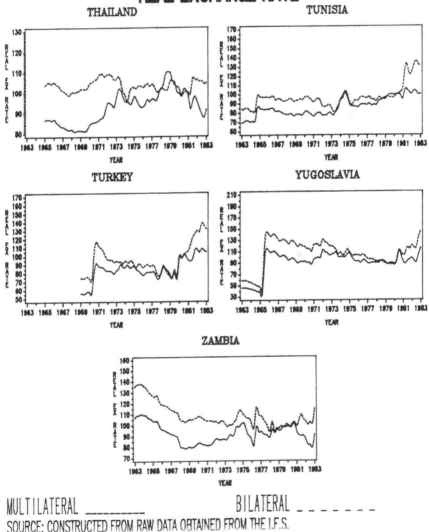

MULTILATERAL AND BILATERAL
REAL EXCHANGE RATE

MULTILATERAL _____   BILATERAL _ _ _ _ _ _

SOURCE: CONSTRUCTED FROM RAW DATA OBTAINED FROM THE I.F.S.

appears that these indexes have not had significant long-term trends during the whole period under consideration. For the shorter more recent periods, however, negative trends can be detected in a number of cases.

Tables 3.2 and 3.3 contain data on the main statistical properties of the multilateral real exchange rate index MRER1 for our 33 countries. These indicators have been calculated for two time periods: 1965-71 and 1972-85. The years 1965-71 correspond to the last years of the Bretton Woods period, during which a majority of countries were pegged to the U.S. dollar. The last period, 1972-85, corresponds to the post-Bretton Woods era, a period during which most advanced countries adopted a dirty (or managed) floating nominal exchange rate system and most of the developing nations maintained some kind of peg. The more important findings that emerge from these tables can be summarized as follows. First, as the diagrams suggested, real exchange rates have been quite volatile in many of these countries, with the extent of this variability being quite different across countries. For example, while in Zambia the difference between the maximum and minimum values of the index for the complete 1965-85 period surpasses 130 points, it was only 25 points in Singapore. A second fact that emerges from Tables 3.2 and 3.3 refers to the increased real exchange rate variability through time. A comparison of the coefficients of variation for 1965-71 and 1972-85 reveals that in all but four countries (Ecuador, Philippines, Turkey and Yugoslavia), the multilateral real exchange rate has been significantly more volatile during the post-Bretton Woods era.

In all of these countries a steady depreciation of the multilateral real exchange rate was observed until a certain date--usually the late 1970s--and a number of these countries have pegged or managed their currency against the U.S. dollar; as the U.S. dollar appreciated in the first part of the 1980s, so did these countries' real exchange rates. Also identifiable is a group of countries whose RERs have exhibited clear long-term trends: Cyprus, India, Malta, Mauritius, Tunisia and Turkey have a definitively strong positive trend (i.e., the RER has depreciated through time), while Bolivia, Ecuador, Ethiopia, Paraguay and Zambia have exhibited a negative (real appreciation) long-run trend. Finally, there are those countries whose RERs do not show a significant long-term trend. However, in spite of the absence of a long-term trend, in some of these cases, Kenya and Mexico, for example, there have been some fairly abrupt jumps in RERs, usually as a result of major nominal devaluations. The degree of RER instability across these countries also has been fairly different, with

TABLE 3.2. Basic Statistical Properties of Multilateral Real
Exchange

Rate Index MRER1 (Quarterly Data 1965-1971)

	Mean	Standard Deviation	Coefficient of Variation	Maximum	Minimum
Bolivia	78.95	5.79	7.33	90.62	72.05
Brazil	62.73	9.05	14.43	92.37	50.02
Colombia	90.93	8.39	9.23	101.59	68.26
Cyprus	76.98	1.58	2.05	80.26	74.34
Dom. Rp.	102.35	2.82	2.75	108.43	97.67
Ecuador	92.17	7.36	10.15	110.19	82.69
El Salva.	97.31	3.37	3.46	106.49	92.30
Ethiopia	108.06	2.82	2.60	115.09	102.51
Greece	86.66	2.34	2.70	92.53	83.65
Guatemala	87.48	2.63	3.00	94.02	83.13
Guyana	68.64	4.32	6.30	75.59	62.46
Honduras	81.11	1.61	1.98	85.14	78.03
India	66.90	6.77	10.12	76.53	52.45
Israel	78.10	9.12	11.68	88.88	65.72
Kenya	91.94	2.85	3.10	97.83	87.60
Korea	81.57	3.98	4.87	88.29	77.12
Malaysia	80.43	2.43	3.02	86.18	76.48
Malta	68.81	4.49	6.53	74.88	62.31
Mauritius	84.61	3.79	4.48	92.65	80.91
Mexico	96.82	2.00	2.06	100.04	51.09
Pakistan	53.31	2.32	4.35	58.12	-
Paraguay	84.53	3.63	4.29	91.40	56.82
Peru	64.95	4.20	6.47	72.29	56.39
Philip.	78.78	11.65	14.79	102.45	68.40
Singapore	88.28	3.04	3.45	94.47	83.89
S. Africa	83.37	2.65	3.18	88.03	80.08
Sri Lanka	29.82	2.11	7.10	32.91	26.22
Thailand	84.45	2.67	3.16	91.09	80.99
Tunisia	80.74	2.68	3.32	85.23	76.41
Turkey	71.88	15.78	21.95	92.35	55.68
Yugoslavia	94.77	17.87	18.85	115.91	35.65
Zambia	87.72	8.25	9.41	104.05	78.96

Source:See text.

TABLE 3.3.   Basic Statistical Properties of Multilateral Real
             Exchange

Rate Index MRER1 (Quarterly Data 1972-1985)

	Mean	Standard Deviation	Coefficient of Variation	Maximum	Minimum
Bolivia	91.00	17.99	19.76	122.92	47.17
Brazil	82.38	16.30	19.79	115.23	61.85
Chile	93.47	26.44	28.29	147.66	18.67
Colombia	103.15	11.12	10.78	124.12	84.71
Cyprus	93.57	8.68	9.27	103.40	74.73
Dom.Rp.	100.57	17.68	17.58	175.51	59.95
Ecuador	98.86	7.93	8.03	114.38	79.60
El Salv.	95.71	21.25	22.20	123.94	51.07
Ethiopia	100.09	17.65	17.63	140.18	64.27
Greece	98.35	4.03	4.10	112.29	90.38
Guatemala	94.60	7.85	8.30	108.46	67.93
Guyana	89.38	12.15	13.59	105.08	63.03
Honduras	92.94	9.47	10.79	106.10	74.58
India	90.93	11.01	12.10	105.30	70.44
Israel	89.42	11.39	12.74	110.68	67.91
Kenya	104.49	6.04	5.78	118.96	92.83
Korea	98.92	6.54	6.61	119.72	87.02
Malaysia	86.80	6.76	7.79	101.72	75.49
Malta	93.12	8.40	9.02	106.02	73.43
Mauritius	97.23	6.17	6.34	111.62	89.15
Mexico	106.72	15.70	14.72	148.95	85.06
Pakistan	97.59	10.81	11.07	125.48	54.81
Paraguay	90.81	17.43	19.19	131.60	56.82
Peru	87.07	16.40	18.83	117.96	61.64
Philip.	101.61	6.43	6.33	123.87	87.99
Singapore	88.68	6.41	7.23	100.99	75.84
S. Africa	98.91	8.26	8.35	116.21	84.50
Sri Lanka	72.64	28.11	38.70	105.63	27.41
Thailand	96.42	5.77	5.98	110.21	84.85
Tunisia	95.32	8.87	9.30	107.50	77.51
Turkey	92.70	13.39	14.45	123.08	73.11
Yugoslav.	103.25	13.94	13.50	133.93	87.76
Zambia	95.43	17.32	18.15	213.73	79.12

Source: See text.

Kenya, for example, being quite stable, while Mexico has exhibited a fair amount of instability.

To investigate RER behavior further, linear trend regressions were estimated for four time periods: 1965-85, 1965-71, 1972-85 and the more recent period 1978-85. For most countries the absolute value of the estimated coefficients for the whole period were small, although in most cases they were significant. A comparison of the number of negative signs of the trend coefficients in the earlier Bretton Woods era and the more recent period shows that during 1965-71 in only 8 of the 33 countries the trend coefficient was small but negative, indicating a weak tendency towards appreciation. However, during 1978-85 in 23 of 33 countries the trend coefficient was negative, and in some cases like Ecuador, Paraguay and Bolivia, fairly large.

4.2. Parallel Markets and RER Behavior: The Cross-Country Evidence

The RER indexes used in the analysis of subsection 4.1 were constructed using data on official nominal exchange rates. However, as pointed out in Section 4.3, in many developing countries at different points in time there have been quite significant parallel (or black) markets for foreign exchange. The coverage and importance of these parallel markets vary from country to country and from period to period. In some cases they are quite thin and used mainly by those nationals wanting to spend vacations abroad who are allowed only a limited quota of foreign exchange at the official rates. In other cases, coverage of the parallel market is very broad and the parallel market exchange rate is the relevant marginal rate for most transactions. The degree of legality of these parallel markets also varies from case to case. While in some cases they are quasi-legal and accepted by the authorities as a minor nuisance, in others they are strongly repressed, with authorities severely persecuting those engaging in black market transactions.

The very nature of these markets--illegal or quasi-illegal--makes it difficult to obtain accurate data on their volume of transactions and their relative importance. However, there are relatively reliable data on parallel market quotations and parallel market premia. Generally speaking, the parallel market premium will become higher as exchange controls become more pervasive and generalized and as fewer transactions are allowed through the official market. In fact, under conditions of generalized exchange controls and rationing, RER indexes computed using official rates will become more and more irrelevant for a number of transactions and in particular for imports.

There is no reason why the parallel market RER index (PMRER) should move closely with the indexes constructed using the official nominal exchange rates. In fact, there are a number of circumstances under which, in a country with pegged nominal official rates, these two RER indexes will tend to move in opposite directions. This will happen, for example, when there is massive domestic credit creation under generalized exchange controls and active parallel markets. Under these circumstances, the higher growth of domestic credit will simultaneously generate an appreciation of the official RER index and a depreciation of the parallel market RER. To investigate this issue further, coefficients of correlation between the parallel market RER and official RER bilateral indexes were calculated [see Edwards (1989a)]. The parallel market index was constructed as:

$$PMRER1_t = (PM)_t \frac{WPI^{US}}{CPI}$$

where $(PM)_t$ is an index of the parallel market bilateral nominal exchange rate with respect to the U.S. dollar, WPI^{US} is the U.S. wholesale price index, and CPI is the domestic country consumer price index. PMRER1, then, is the bilateral parallel index equivalent to BRER1 in Section 4.1. The results obtained clearly capture the fact that the parallel and official RER indexes indeed behave very differently. In fact, in 13 of the 28 cases the coefficients of correlation were negative!

5. REAL EXCHANGE RATE VOLATILITY AND ECONOMIC PERFORMANCE[13]

An important finding about RER behavior reported in the previous section refers to the increased degree of volatility experienced by RERs. From a theoretical perspective, it has been well established that real exchange rate disequilibrium and heightened uncertainty regarding RER behavior will have negative effects on economic performance (Willet 1986). Empirically, however, there has been more difficulty finding evidence that supports these theoretical insights. According to the IMF (1984), for example, there is insufficient evidence linking increased real exchange rate instability to less active international trade. To quote (IMF 1984, page 36):

The large majority of empirical studies on the impact of exchange rate variability on the volume of international trade are unable to establish a systematic significant link between measured exchange rate variability and the volume of international trade, whether on an aggregate or on a bilateral basis.

In its own empirical investigation the IMF found no empirical evidence of a relationship between exchange rate instability and bilateral trade flows for seven industrial nations. Cavallo and Cotani (1985), however, found some evidence of a negative relation between RER variability and economic performance. Their analysis looked only at bilateral rates and concentrated on official nominal exchange rates. We have seen above, however, that not only do effective and bilateral rates behave in a significantly different way, but also black market and official rates often move in opposite directions.

In this section we report some regression results on the relation between real exchange rate instability and economic performance using cross section data for 23 of the 33 countries of the previous section.[14] The data were broken into two periods: 1965-71, corresponding to the last six years of the Bretton Woods System and 1978-85, corresponding to the most recent period with an international floating exchange rate system. The analysis dealt with four different measures of economic performance: (1) average rate of growth of real GDP over each of the two periods considered; (2) average rate of growth of real GDP per capita; (3) average rate of growth of real export; and (4) average investment-output ratio. The regressions were estimated in both levels and logs. The structural form of the equations actually estimated were:

$$X_n = \alpha + \beta \, \delta_n + \Sigma \, j_i \, Z_{ni} + \epsilon_n \tag{16}$$

and

$$\log X_n = \alpha' + \beta' \log \delta_n + \Sigma \, j_i \log Z_{ni} + \epsilon_n \tag{17}$$

where the following notation has been used:

X_n —   performance variable (average growth of GDP; average growth of exports; and average investment ratio; for country n.

δ_n = coefficient of variation of the real exchange rate index for country n. Three indexes were considered: bilateral; effective and black market.

Z_{in} = other relevant variables.

e_n, x_n = error terms, assumed to have the usual properties.

Depending on the left-hand-side variable, different Z_{in}'s were included. In the two output growth equations, two Z_{in}'s (i.e., "other" variables) were incorporated in the regressions: investment-output ratio and variability (coefficient of variation) of the international terms of trade. The investment variable was incorporated as a way to capture the effects of capital accumulation in explaining cross-country growth differentials. Its coefficient is expected to be positive. The terms of trade variability was added in an effort to incorporate other sources of external instability faced by these nations, and its sign is expected to be negative. In the growth of exports equations the variability index of the terms of trade was the only Z_{in} included; its sign is expected to be negative. Finally, in the investment ratio equation no additional variables (Z_{in}'s) were incorporated.

Our interest is to find out whether greater real exchange rate instability has indeed been associated with some kind of "poorer" economic performance. In terms of equations (15) and (16), we are interested in testing whether the coefficients ß and ß' are significantly negative or if, as suggested by the IMF, there has been no relation between exchange rate instability and economic performance. The output growth equations performed better when they were estimated in levels. The exports and investment equations, on the other hand, generated better fits when estimated in logs. To save space, we only report the better results. Other estimates, including those obtained when nonlinear terms were added, are available from the author on request.

Tables 3.4 through 3.6 contain the regression results obtained when "official" RER indexes were used in computing the RER variability measure. A number of interesting results emerge from these tables. In particular, there seems to be a definite difference in the way these variables interacted during the Bretton Woods and the floating rates periods. The results quite strongly suggest a structural break between the two periods. While real exchange rate instability played no role in the Bretton Woods era, it helps explain cross-country differentials in economic performance during the more recent period. Moreover, during the more recent floating rates era, there is a quite clear

TABLE 3.4. Cross-Country Regressions: Average Growth of Real
GDP and Real Exchange Rate Variability, 1965-71
and 1978-85 (OLS)

| | | Var. of RER | | Investment | Var. | |
Period	Constant	Bilateral	Effective	Ratio	T of T	R^2
65-71	1.716	-0.091	-	0.303	-0.068	0.306
	(0.600)	(-0.747)		(2.436)	(-0.457)	
65-71	1.460	-	-0.086	0.310	-0.056	0.302
	(0.519)		(-0.681)	(2.428)	(-0.383)	
78-85	1.898	-0.185	-	0.194	-0.043	0.579
	(1.036)	(-3.044)		(3.261)	(-0.895)	
78-85	2.244	-	-0.279	0.157	-0.016	0.532
	(1.120)		(-2.593)	(2.500)	(-0.300)	

Numbers in parentheses are t-statistics.

TABLE 3.5. Cross-Country Results: Investment Ratio and Real
Exchange Rate Variability, 1965-71 and 1972-85
(OLS)[a]

| Period | Constant | Log of RER Variability | | R^2 |
		Bilateral	Effective	
65-71	2.886 (64.113)	0.038 (0.626)	-	0.013
65-71	2.839 (24.464)	-	0.033 (0.507)	0.009
72-85	3.203 (39.849)	-0.158 (-2.092)	-	0.127
72-85	3.472 (16.940)	-	-0.173 (-2.048)	0.123

[a]The dependent variable is the log of the gross investment to
GDP Ratio.

Numbers in parentheses are t-statistics.

TABLE 3.6. Cross-Country Regressions: Export Growth and
RER Variability, 1965-1971 and 1978-85 (OLS)[a]

| Period | Constant | Log of Variability | | Log of Variability | R^2 |
		Bilateral RER	Effective REER	T of T	
65-71	2.067 (1.715)	-	-0.121 (0.296)	-0.664 (-1.487)	0.147
65-71	2.644 (6.488)	-0.333 (-1.615)	-	-0.145 (-0.953)	0.195
78-85	1.353 (2.928)	0.326 (0.898)	-	0.099 (0.205)	0.067
78-85	0.917 (1.207)	-	0.286 (0.789)	0.019 (0.093)	0.054

[a]The dependent variable is the log of the average rate of
growth of real exports.

Numbers in parentheses are t-statistics.

negative relation between real exchange rate instability and our real performance measures.

Table 3.4 contains the regression results for the average rate of growth of real GDP and real GDP per capita. The results are quite satisfactory, especially for the floating rates era. The R^2s indicate that a fairly significant fraction of the variability of average rates of growth across countries can be explained by these equations. Not surprisingly, the investment ratio is positively related to the average rates of growth. Those countries that accumulate capital more rapidly generally have grown at a faster than average rate. This result holds for the Bretton Woods period as well as for the floating rates period. Notice, however, that there is a marked difference in the magnitude of the coefficients. For the most recent era, the point estimates are almost half those obtained for the Bretton Woods period. As noted, the coefficients of the RER variability indexes are quite different across periods. The results do provide quite strong support for the hypothesis that during the floating rates period higher real exchange rate instability has been associated with poorer economic performance. It is interesting to note that the coefficients of the index of variability of the real effective rate are higher than those of the index of variability for the real bilateral rate. In all equations the index of instability of external terms of trade was not significantly different from zero.

How can we account for the differences in the results for the Bretton Woods and floating rates periods? A possible explanation is the different nature of real exchange rate instability in both periods. Under the old institutional arrangements, real exchange rate movements were much more predictable, since the institutional framework ruled out wide daily fluctuations in third party bilateral exchange rates.

Table 3.5 presents the regression results for the investment equation. A double log specification was used. In many ways these results are similar to those on real output growth, indicating that during the floating rate period there has been a fairly strong negative relation between the degree of real exchange rate instability and investment. Notice, however, that these equations explain only a very low percentage of the cross-country variations of investment ratios. Finally, Table 3.6 contains the results for the export growth equations. Interestingly enough, these results are in line with those of the IMF, indicating that there is no significant connection between RER instability and exports performance. However, a word of caution related to measurement is needed here. The export growth data used in those regressions refer to exports in real U.S. dollars, and have been computed as the rate of growth of exports in U.S. dollars deflated by the U.S. WPI. There is, then, a potential valuation problem that may be responsible for the coefficients of the variability indexes being nonsignificant.

The results reported above were obtained using indexes of instability of RER that were constructed using official data on nominal exchange rates. However, as was shown in Section 4, in developing countries there often are significant departures of the black market rate from official rates. For this reason, performance equations using instability of black market rates also were estimated. The following result was obtained for the floating rates period and for the rate of growth of real GDP.

```
Growth Real GDP    = 2.507   -     0.190 Variability BMRER
                     (1.812)     (-3.445)

    + 0.166 Investment Ratio  -  0.004 Variability T of T
      (2.939)                    (-0.910)
```

$$R^2 = 0.646$$

This result supports our previous findings, indicating that higher real exchange rate instability clearly has been associated with poorer economic performance (i.e., real growth) during the Bretton Woods period. The interesting thing here is that since we dealing with black market instability, this measure clearly is dependent on the policies followed by the government. To analyze these issues further, regressions including both official RER instability and variability of the black market RER were estimated. This allows us to have some idea of how, other things given, each of these sources of instability has affected economic performance:

```
Growth Real GDP    = 3.208   -     0.711 Variability RER
                     (1.767)     (-1.427)

    + 0.153 Investment Ratio  +  0.009 Variability T of T
      (2.767)                     (0.176)

    - 0.138 Variability BMRER
      (-2.239)
```

$$R^2 = 0.694$$

6. CONCLUDING REMARKS

In this paper I analyzed some of the most important aspects related to the concept and measurement of real exchange rates. In addition I provided an empirical analysis of real exchange rate behavior in a large group of developing countries. The discussion emphasized that it is crucially important to make a distinction between equilibrium and disequilibrium changes of real exchange rates. For this reason I presented an intertemporal model of equilibrium real exchange rate behavior in a fully optimizing economy and discussed how alternative macroeconomic policies can result in real exchange rate misalignment.

The empirical analysis showed that in the recent years bilateral and multilateral real exchange rates have exhibited markedly different behavior, an indication that ignoring those problems emerging from the existence of a floating rate international monetary system can result in greatly biased policy recommendations. It has also been pointed out that in a large number of developing countries parallel markets for foreign exchange can be quite important. Thus using RER constructed with official nominal rates can also result in misleading conclusions.

Finally, the effects of real exchange rate variability on economic development in a group of developing countries is empirically investigated. Using a cross-section data set, it was found that higher RER volatility has been associated with lower output growth and lower investment. There are no indications, however, that higher variability in the RER affects the level of exports.

NOTES

1. Financial support from the National Science Foundation and UCLA's Academic Senate is gratefully acknowledged. David Gould provided able research assistance. As always, I am indebted to Miguel Savastano for his kind assistance.

2. The material presented in this paper summarizes and expands work I have been doing on real exchange rates in developing countries during the last six years. See the references for a list of related papers.

3. See, for example, Dornbusch (1982) and IMF (1984). See Edwards and Ng (1985) for an exhaustive review of alternative definitions of the real exchange rate.

4. See, for example, Krueger (1983, p. 18), McKinnon (1979, pp. 121-28), Officer (1982, pp. 148-50) and Williamson (1983b, p. 14). For discussions on the merits and problems of the PPP theory see, for example, Officer (1982) and Frenkel (1978, 1981).

5. Notice that this definition assumes that the law of one price holds for tradable goods. This, of course, is a debatable issue.

6. On this type of model see, for example, Dornbusch (1980) and Mussa (1984). Notice that since, as explained below, the real exchange rate defined as above does not have to move in the same direction, then the PPP-defined real exchange rate (an improvement of the latter) does not necessarily result in an improvement of the current account.

7. For recent discussions on the causes and magnitudes of disequilibrium or misalignments of the real exchange rates, see, for example, Dornbusch (1982), Williamson (1983b) and McKinnon (1984).
   A common confusion in the literature is to use the concepts of the real exchange rate and the terms of trade interchangeably. See, for example, Isard (1983). Of course, since the terms of trade are defined as the relative price of exportables to importables, and the real exchange rate is defined as the relative price of tradables to nontradable goods, there is no reason for them to be equivalent. In fact, as will be discussed below (Section 2.3), there are circumstances

where these two variables will tend to move in the opposite direction. Williamson (1983b) has recently stressed the importance of distinguishing between the terms of trade and the real exchange rate. Katseli (1984) has recently shown, using a cross-country data set, that these two variables have tended to behave quite differently in recent years.

8. What we have called here the long-run sustainable equilibrium real exchange rate is (somewhat) equivalent to Williamson's (1983b) "fundamental equilibrium real exchange rate."

9. Edwards (1989a) develops a formal model of exchange rate misalignment and balance-of-payments crises.

10. Some authors have also suggested using an alternative indicator of competitiveness constructed as the ratio of export unit cost to import unit costs. Of course, the problem with this is that it confuses the terms of trade with the real exchange rate! See Williamson (1983b). See also Connolly and Lackey's (1983) proposition for using the "real monetary parity."

11. This criticism implicitly assumes that the "law of one price" holds for homogeneous tradable goods. See, however, Kravis and Lipsey (1983) and Isard (1977).

12. On the selection of the "appropriate" weights, see Branson and de Macedo (1982) and Branson and Katseli (1982).

13. This section draws partially on Edwards (1989d).

14. These are the only countries for which there are data on all the required variables.

REFERENCES

Artus, J. (1978). "Methods of Assessing the Long-Run Equilibrium Value of an Exchange Rate," Journal of International Economics 8: 277-99.

Artus, J. and M. Knight (1984). "Issues in the Assessment of the Exchange Rate in Industrial Countries." IMF Occasional Paper 29, Washington, DC.

Barro, R. (1983). "Real Determinants of the Real Exchange Rate," unpublished ms., University of Chicago.

Branson, W. and L. Katseli (1982). "Currency Baskets and Real Effective Exchange Rates," in M. Gersovitz, et al. (eds.), The Theory and Experience of Economic Development. London: Allen & Unwin.

Branson, W. and J. de Macedo (1982). "The Optimal Choice of Indicators for a Crawling Peg," Journal of International Money and Finance 2: 1965-78.

Bruno, M. (1982). "Import Competition and Macroeconomic Adjustment Under Wage-Price Rigidity," in J. Bhagwati (ed.), Import Competition and Response. Chicago: University of Chicago Press.

Calvo, G. (1986). "Fractured Liberalism: Argentina Under Martinez de Hoz," Economic Development and Cultural Change 34: 511-34.

Cavallo, D. and J. Cotani (1985). "Real Exchange Rates in LDC's," unpublished ms., Washington: World Bank.

Cline, W. (1989). "Latin American Debt: Progress, Prospects and Policy," in S. Edwards and F. Larrain (eds.), Debt, Adjustment and Recovery. London: Basil Blackwell.

Connolly, M. and C. Lackey (1983). "A Test of the Exchange Market Pressure Model, Mexico 1955-1982," working paper, University of South Carolina.

Corbo, V., J. de Melo, and J. Tybout (1986). "What Went Wrong With the Recent Reforms in the Southern Cone," Economic Development and Cultural Change 34: 607-40.

Dervis, K. and P. Petri (1987). "The Macroeconomics of Successful Development: What Are the Lessons?" NBER Macroeconomics Annual, 1987.

de Vries, M. (1968). "Exchange Depreciation in Developing Countries," IMF Staff Papers, 15: 560-602.

Diaz-Alejandro, C. (1983). "Real Exchange Rates and Terms of Trade in the Argentine Republic 1913-1976," in M. Syrquin and S. Teitel (eds.), Trade, Stability, Technology and Equity in Latin America. New York: Academic Press.

Dornbusch, R. (1974). "Tariffs and Nontraded Goods," Journal of International Economics 4: 117-85.

————— (1980). Open Economic Macroeconomics, New York: Basic Books.

————— (1982). "Equilibrium and Disequilibrium Exchange rates," Zeitschrift fur Wirtschats und Sozialwissenshaften 6: 573-99.

Edwards, S. (1984). "The Order of Liberalization of the External Sector," Princeton Essays on International Finance No. 156, Princeton: Princeton University Press.

————— (1989a). Real Exchange Rates, Devaluation and Adjustment: Exchange Rate Policy in Developing Countries. Cambridge, MA: MIT Press.

————— (1989b). "Tariffs, Capital Control and Equilibrium Real Exchange Rates," Canadian Journal of Economics (forthcoming).

————— (1989c). "Temporary Terms of Trade Disturbances, Real Exchange Rates and the Current Account," Economica (forthcoming).

_____ (1989d). "Implications of Alternative International Exchange Rate Arrangements for the Developing Countries," in H. Vosgerau (ed.) New Institutional Arrangements for the World Economy. Berlin: Springer-Verlag.

Edwards, S. and F. Ng (1985). "Trends in Real Exchange Rate Behavior in Selected Developing Countries," CPDTA Working Paper, Washington, D.C.: World Bank.

Frenkel, J. (1978). "Purchasing Power Parity: Doctrinal Perspectives and Evidence From the 1920s," Journal of International Economics 8: 169-91.

_____ (1981). "The Collapse of Purchasing Power Parity During the 1970s,"European Economic Review 70: 145-65.

Frenkel, J. and M. Mussa (1984). "Asset Markets, Exchange Rates and the Balance of Payments," in R. Jones and P. Kenen (eds.), Handbook of International Economics, vol. II. Amsterdam: North-Holland.

Genberg, H. (1978). "Purchasing Power Parity Under Fixed and Flexible Exchange Rates," Journal of International Economics 8: 247-76.

Gulhati, R., et al. (1985). "Exchange Rate Policies in Eastern and Southern Africa: 1965-1983," Staff Working Paper 720, Washington, DC: World Bank.

Harberger, A. (1981). "The Real Exchange Rate of Chile: A Preliminary Survey," paper presented at a Conference on Economic Policy, Vina del Mar, Chile (April).

_____ (1983). "Welfare Consequences of Capital Inflows," paper presented at a World Bank Conference, Washington, D.C. (October).

Hooper, P. and J. Morton (1982). "Fluctuations in the Dollar: A Model of Nominal and Real Exchange Rate Determination," Journal of International Money and Finance 1: 39-51.

Houthakker, H. (1962). "Exchange Rate Adjustment," in Factors Affecting the U.S. Balance of Payments. Washington, DC: U.S. Government Printing Office.

_____ (1963). "Problems of International Finance," Agricultural Policy Review 3: 12-13.

International Monetary Fund (IMF) (1984). "Exchange Rate Volatility and World Trade." Occasional Paper 28, Washington, DC.

Isard, P. (1977). "How Far Can We Push the Law of One Price?" American Economic Review 67: 942-48.

_____ (1983). "An Accounting Framework and Some Issues for Modeling How Exchange Rates Respond to 'News'," in J. Frenkel (ed.) Exchange Rates and International Macroeconomics. Chicago: University of Chicago Press.

Katseli, L. (1984). "Real Exchange Rates in the 1970s," in J. Bilson and R. Marston (eds.), Exchange Rate Theory and Practice. Chicago: University of Chicago Press.

Keynes, J.M. (1924). A Tract on Monetary Reform, reprinted in J.M. Keynes, Collected Writings, Vol. IV. London: Macmillan.

_____ (1930). A Treatise on Money. London: Macmillan.

Kravis, I. and R. Lipsey (1983). "Toward an Explanation of National Price Levels," Princeton Studies in International Finance 52, Princeton: Princeton University Press.

Krueger, A. (1978). Foreign Trade Regimes and Economic Development: Liberalization Attempts and Consequences. Cambridge, MA: Ballinger.

_____ (1983). Exchange Rate Determination. Cambridge: Cambridge University Press.

Maciejewski, P. (1983). "'Real' Effective Exchange Rate Indexes: A Re-Examination of the Major Conceptual and Methodological Issues," IMF Staff Papers 30: 491-541.

McKinnon, R. (1979). Money in International Exchange: The Convertible Currency System. Oxford: Oxford University Press.

_____ (1984). International Standard for Monetary Stabilization. Cambridge, MA: MIT Press.

Mundell, R. (1971). Monetary Theory. Pacific Palisades, CA: Goodyear.

Mussa, M. (1979). "Empirical Regularities in the Behavior of Exchange Rates and Theories of the Foreign Exchange Market," Carnegie Rochester Conference Series on Public Policy.

_____ (1984). "The Theory of Exchange Rate Determination," in J. Bilson and R. Marston (eds.) Exchange Rate Theory and Practice. Chicago: University of Chicago Press.

Neary, P. (1988) "Determinants of the Equilibrium Real Exchange Rate," American Economic Review, (March): 210-15.

Neary, P. and D. Purvis (1983). "Real Adjustment and Exchange Rate Dynamics," in J. Frenkel (ed.), Exchange Rates and International Macroeconomics. Chicago: University of Chicago Press.

Officer, L. (1976). "The Productivity Bias in Purchasing Power Parity: An Econometric Investigation," IMF Staff Papers 23: 545-79.

_____ (1982). Purchasing Power Parity and Exchange Rates: Theory, Evidence and Relevance. Contemporary Studies in Economic and Financial Analysis No. 35 Greenwich, CT: JAI Press.

Viner, J. (1937). Studies in the Theory of International Trade. Clifton, NJ: Kelley Publishers.

Willet, T. (1986). "Exchange Rate Volatility, International Trade and Resource Allocation: A Perspective on Recent Research," Journal of International Money and Finance, Supplement (March): 101-12.

Williamson, J. (1983a). The Open Economy and the World Economy. New York: Basic Books.

_____ (1983b).   The Exchange Rate System.   Institute for International Economics, Cambridge, MA: MIT Press.

4

Empirically Testing the Law of One Price in International Commodity Markets: A Rational Expectations Approach

Barry K. Goodwin

1. INTRODUCTION

The law of one price (LOP) is an important ingredient in theories of international trade and exchange rate determination. As Officer (1986) notes, without the imposition of this principle there would not even be the traditional "pure" theory of international trade. Similarly, much of the "monetary theory" of trade would have to be reconsidered. In short, the LOP maintains that the foreign price of a commodity, once adjusted for exchange rates and transportation costs, will be equal to the domestic price of the commodity. This equality will be established and maintained by the profit-seeking actions of international commodity traders and arbitragers.

The law of one price has little to do with money (Niehans 1984). In a barter world, if transportation costs and other trade impediments are abstracted from, the price of a commodity in terms of real goods clearly must be the same everywhere in the trading area, both nationally and internationally. More generally, there can be only one equilibrium set of barter terms of trade. If this were violated, traders could make unlimited and riskless profits by buying a commodity where it is relatively cheap and selling it where it is dear. These ideas are easily extended to consider the case of international trade between monetary economies. The implication is that domestic prices, once converted to a common currency, will be the same as international prices.

The international trade literature contains a large volume of theoretical and empirical investigations of the law of one price. A review and assessment of many recent considerations of the LOP is contained in Officer (1988). These investigations are usually conducted on a more general level as tests of the purchasing power parity (PPP) theory using aggregate data and price indexes. The results differ as to degree, but most evidence is in favor of rejecting strict adherence to the purchasing power parity theory. Similarly, empirical results regarding price parity among disaggregated commodities generally contradict the LOP theory.

Theories regarding purchasing power parity are usually thought to have originated in the work of Cassel (1918). However, there are much earlier allusions to the ideas of parity in international prices. Officer (1982) attributes the first statement of purchasing power parity to Spanish scholars of the 16th century, the Salamanca School. References to parity are also found throughout the works of the bullionist period and later in Mill (1929) and Marshall (1926).

The results of many recent investigations have indicated that deviations from parity are common in the short run, but that parity is often found to hold in the long run. Kravis and Lipsey (1978) conclude that markets work in the "textbook fashion" but slowly rather than instantaneously and thus that it is "unlikely that a high degree of national and international commodity arbitrage is typical of the real world." Isard (1977) argues that exchange rate changes substantially alter the dollar-equivalent prices for most narrowly defined goods and that these relative price effects seem to persist for at least several years. Genberg (1978) finds that adjustments to the real exchange rate may have lags of up to seven years and thus concludes that purchasing power parity is mainly a longer-run phenomenon. Equivalent conclusions are reached by Richardson (1978), Frenkel (1981), Frenkel and Johnson (1976), Gailliot (1970), Dornbusch (1976), and Wihlborg (1979).

However, conflicting results have been obtained in other work. Hodgson and Phelps (1975) argue that the major impact of price level movements on exchange rates occurs within three months and thus that the purchasing power parity theory may offer a useful explanation of exchange rate behavior over shorter periods than those to which it generally has been applied. Rogalski and Vinso (1977) have reached conclusions consistent with the purchasing power parity theory by finding that freely floating exchange rates react immediately or nearly so to changes in relative inflation rates. Officer (1986), employing a model making use of the tradable/nontradable goods dichotomy, obtains results that offer support for the law of one price at an aggregate level.

Less attention has been directed to specific investigations of the law of one price using prices of basic homogeneous commodities. Results here are similar to the more general conclusions regarding the purchasing power parity theory. In general, the evidence indicates that disaggregated commodity arbitrage takes place, but that short-run violations of the LOP are often observed and that adherence is much more likely to occur in the long run. Crouhy-Veyrac et al. (1982)

attribute such deviations to transfer costs. Protopapadakis and Stoll (1983) also point to transport costs and other impediments to commodity arbitrage as reasons for short-run failure of the LOP. Protopapadakis and Stoll (1986) formulate a multi-equation model that distinguishes between short-run and long-run prices. Their results indicate that the LOP almost never holds in the short run, but that the long-run version of the LOP receives strong empirical support. An equivalent conclusion is reached by Jain (1980) in an analysis of price parity in commodity futures markets. Jabara and Schwartz (1987) also find that disaggregated agricultural commodity prices commonly violate the LOP.

The question of adherence to the law of one price in international commodity markets has important implications for the appropriate approach to theoretical and empirical modeling of trade. Chambers and Just (1979) critically note that most analyses of exchange rates and international trade explicitly assume adherence to the law of one price in absolute terms. However, the empirical evidence would seem to question the validity of such assumptions. Models relying upon such suppositions may be misspecified and may therefore yield incorrect results and invalid implications.

An important shortcoming of the existing empirical literature addressing the LOP question is that such analyses typically have assumed that parity should hold contemporaneously. This approach overlooks the fact that international commodity arbitrage and trade take place over time as well as across spatially separated markets. It takes time to move goods from one market to another. Recognizing this fact, we should not expect to see parity holding for contemporaneous spot prices unless arbitragers have perfect foresight or unless prices are constant. A more reasonable approach would be to expect international commodity arbitragers to act upon the price that they expect to prevail in the market in which they will sell when their goods are delivered. Thus, we would instead expect to see parity holding for expected prices.

The objective of this paper is to incorporate explicitly the role of expectations into an empirical investigation of the law of one price. The intertemporal elements of the trade and arbitrage process suggest that the law of one price will receive stronger empirical support when the role of price and exchange rate expectations is given explicit consideration. In contrast, the conventional approach is to ignore the role of expectations while using ex-post contemporaneous price data to test for parity. The application is to international markets for primary homogeneous commodities. The investigation makes use of the generalized method of moments (GMM) procedures set forth by

Hansen (1982) and Hansen and Singleton (1983). The econometric procedures apply instrumental variables estimation techniques directly to orthogonality conditions, which are implied by the first-order conditions of international commodity arbitragers' stochastic intertemporal optimization problems. The resulting estimates provide a rational expectations interpretation of the law of one price that can be compared to the standard formulations of the LOP, which utilize ex-post contemporaneous prices.

The paper proceeds according to the following plan. The next section develops a model of the law of one price in international commodity markets. An econometric methodology aimed at incorporating the role of price expectations in empirical tests of the LOP is also introduced. The third section applies the models and methodology to an empirical consideration of the LOP in international markets for several important U.S. agricultural commodities. The fourth section extends the analysis to a consideration of cross-currency trade and international futures markets. The fifth section contains a discussion of the role of transportation and interest costs in questions of price parity in international commodity markets. The final section contains a brief review of the results and offers some concluding remarks.

2. A MODEL OF THE LAW OF ONE PRICE

Basic tests of the law of one price are best carried out using prices of basic (non-manufactured or non-processed) commodities. There are three basic reasons for the superiority of this approach. First, following the hedonic pricing proposal of Rosen (1974), a good's price may reflect the presence and quality of certain utility-bearing attributes. In this case, we would expect goods bearing different attributes, that is differentiated goods, to have different values and thus different prices. A basic primary commodity is most likely to possess identical attributes regardless of origin or destination. Second, aggregated data carry with them the problems associated with indexes and aggregation measurement errors. Tests that use aggregate data to reject the law of one price may fail in part because of aggregation and index construction errors. Finally, when tests of the LOP are carried out using highly aggregated price data, there is a high probability that exchange rates will be endogenous to the system used for testing. The existence of such endogeneity explicitly biases the results of standard regression tests. There is a much smaller chance of exchange rates being endogenous when the tests are carried out using basic commodity prices.

Most empirical tests of the law of one price utilize a model similar to that of Richardson:

$$P_{it} = \alpha_0 \, P_{it}^{*\,\alpha_1} \, \pi_{12t}^{\alpha_2} \, T_{it}^{\alpha_3} \, R_{it}^{\alpha_4} \tag{1}$$

where:
P_{it} = country one's price of commodity i in time t,

P_{it}^{*} = country two's price of commodity i in time t,

π_{12t} = rate of exchange for currency two in terms of currency one,

T_{it} = transfer and transactions costs of trade in commodity i between countries one and two,

R_{it} = residual reasons for price differences between countries one and two, and

α_0, α_1, α_2, α_3, and α_4 are parameters.

Strict adherence to the law of one price requires that the domestic price of a good, once adjusted for exchange rates, transfer costs, and any differences in quality, will be equal to the foreign price of the good. Should a disparity between these prices be detected by international commodity arbitragers, they will actively seek profits by buying the good in the lower priced market and transferring it to the higher priced market, selling it there. These actions will continue until prices are equalized. Thus, for a basic homogeneous commodity, adherence to the LOP requires that:

$$\alpha_0 = \alpha_1 = \alpha_2 = \alpha_3 = 1, \text{ and } \alpha_4 = 0 \tag{2}$$

Thus, equation (1) becomes a statement of the law of one price.

There are several weaknesses inherent in the standard approach to testing the law of one price. First, a major drawback to this approach exists because independent information about transportation costs is rarely available. This problem commonly has been handled by assuming transport costs to be constant over the period of study. If transport costs were truly constant, then T_{it} could be removed as a variable in the equation, since the analysis is being conducted in a regression framework. Another approach to the problem is to assume that transport costs can be approximately represented as a constant proportion of nominal commodity prices. In this case, the influences of transport costs on commodity prices are reflected in the multiplicative constant term, α_0, which is no longer required to be equal to one.

For the preliminary empirical analyses conducted in this paper, the assumption of constantly proportional transfer costs is maintained. However, this assumption will be relaxed in a later segment of the investigation. It is also assumed that, since basic homogeneous commodities are the focus of this analysis, there are no residual reasons for price differences. Thus, R_{it} is omitted and treated as an unobservable random disturbance.

Another serious drawback to the standard approach to testing price parity is that regression tests of the form given by equation (1) necessarily require that one of the commodity prices be taken to be exogenous. For any individual commodity, this is incorrect. While exchange rates often can be assumed to be exogenous to a particular commodity price, prices in two trading countries are simultaneously determined regardless of the relative sizes of the countries (Protopapadakis and Stoll, 1983). This is because information is shared across markets and because agents operate in more than one market at a time. The existence of such simultaneity biases the resulting parameter estimates and thereby confounds attempts at statistical inference. A methodology aimed at overcoming this weakness is applied in the reformulated versions of the LOP pursued in this analysis.

A final shortcoming associated with the standard approach to testing the law of one price is that contemporaneous ex-post realized domestic and foreign prices are utilized in the empirical estimation and testing. This analysis will instead utilize a simple model that explicitly considers the role of expectations. We will assume that each market specializes in either importing or exporting the commodity in question. This assumption draws support from traditional trade theory which asserts that countries will specialize in producing and exporting those commodities for which they have a comparative advantage. For price comparisons between corresponding exporting and importing markets, we will assume that the home country is primarily an exporter and the foreign country is primarily an importer of the commodity in question. Note that this merely provides a rule for defining the domestic country. In this case, the consideration of expectations is limited to one side of the exchange. Namely, it is assumed that exporters respond to their expectations of prices at the time of delivery in the foreign market and that this in turn influences the price of the commodity in the domestic market. The result is parity between current domestic prices and expected future foreign prices. Alternatively, for comparisons between two importing markets, expectations are relevant to both sides of the exchange. In particular, we will assume that importing markets are linked through a common exporting market. If delivery lags between the markets are equal, the relevant price comparisons are of a contemporaneous nature, involving ex-ante expected prices in each of the importing markets.

Under the assumptions outlined above, the basic tests of the law of one price pursued in this analysis are of the form specified by the following equations:

$$P_{it} = a_0 \, (P_{it}^{*})^{a_1} \, (P_{12t})^{a_2} \tag{3}$$

$$P_{it} = \beta_0 \, (E_t\{(P_{it+j}^{*})^{b_1} \, (P_{12t+j})^{b_2}\}) \tag{4}$$

$$E_t\{P_{it+j}\} = c_0 \, (E_t\{(P_{it+j}^{*})^{s_1} \, (P_{12t+j})^{s_2}\}) \tag{5}$$

where $E_t\{x_{t+j}\}$ is the mathematical expectation of the value of x in time $t+j$, conditional on information available in time t. Equation (3) represents the standard model of the LOP; equation (4) represents the expectations-augmented model of the LOP for price relationships between an exporting and a corresponding importing market, and equation (5) represents the expectations-augmented model of the LOP for price relationships between two net importing markets. The LOP price is rejected for the respective models when the parameters α_1 and α_2, β_1 and β_2, and r_1 and r_2 are jointly found to be significantly different from one.

While the conditions for adherence to the law of one price, namely commodity price parity, are intuitively obvious, it is instructive to consider a simple intertemporal model that yields these conditions as a result of arbitrage and trade profit maximization. The arbitrager's problem is to undertake trade over time, maximizing expected profits given a stream of exogenous commodity prices.

Consider a representative agent engaged in arbitrage and trade activities for a single commodity in two spatially separated markets. For the sake of simplicity, we will assume that agents do not have the option of storage. This assumption does not alter any of the analytic results and eliminates the need to consider storage and basis in any specific terms. The agent acts to maximize expected profits by choosing quantities of a commodity to sell in the domestic market in time t and to export to the foreign market for sale upon delivery in time $t+j$. The delivery lag of j periods arises because the physical transfer of commodities in the trade process necessarily takes place over time. The agent faces certain costs associated with trade and arbitrage activities. It is useful to consider these costs as being comprised of two separable components. First and most obvious are those costs associated with transporting the commodity from the domestic market to the foreign market. For the present, it is assumed that such transportation costs can be approximately represented as a constant proportion of the domestic commodity price. A second component consists of those arbitrage costs exclusive of transportation charges. These costs include set-up costs, handling charges, acquisition expenses, and the costs of inputs into the arbitrage and trade process. It is assumed that this component of costs varies directly with the total amount of the commodity placed into arbitrage and trade activities (in both the domestic and foreign markets).

Under these conditions, the agent's problem is to maximize expected profits, given by $E_t\ V(q_t, q_t^*)$, where:

$$V(q_t, q_t^*) = \Sigma_{t=0}^{\infty}\ \{\ \beta^t[p_t q_t + \beta^j \pi_{t+j} p_{t+j}^* q_t^* - C(Q_t) - \alpha p_t q_t^*]\}\quad (6)$$

where: q_t =   quantity sold in the domestic market, time t,

   q_t^* =   quantity exported to the foreign market, time t,

   p_t =   commodity price in the domestic market, time t,

   p_{t+j}^* =   commodity price in the foreign market, time t+j,

   π_{t+j} =   rate of currency exchange, time t+j,

   $C()$ =   commodity arbitrage cost function,

   Q_t =   total quantity of commodity engaged in trade and arbitrage activities, time t (note: $Q_t = q_t + q_t^*$),

   α =   proportional transfer cost parameter,

   β =   a discount factor, $0 < \beta \leq 1$.

Again, note that $E_t(x) = E(x|\Omega_t)$ where E is the mathematical expectations operator conditional on the information set available at time t, Ω_t. We assume that $\Omega_t \subset \{p_t, p_t^*, x_t, \Omega_{t-1}\}$ for all t, where x_t is a vector of random variables exogenous to the agent.

The first-order conditions for profit maximization for all $k \leq 0$ are given by:

$$E_{t+k}\ \{\ \beta^k[p_{t+k} + C'(Q_{t+k})]\ \} = 0 \quad (7)$$

$$E_{t+k}\ \{\ \beta^k[\beta^j \pi_{t+j+k} p_{t+j+k}^* - C'(Q_{t+k}) - \alpha p_{t+k}]\ \} = 0 \quad (8)$$

If we consider these conditions to be adequate representations of aggregate profit-maximizing behavior, we can arrange them at $k=0$ to yield:

$$E_t \{ p_t - \beta^j p_{t+j} p_{t+j}^* + a p_t \} = 0 \qquad (9)$$

which is equivalent to:

$$p_t - \beta_0 E_t \{ p_{t+j} p_{t+j}^* \} = 0 \qquad (10)$$

where $\beta_0 = \beta^j (1 + a)^{-1}$.

Equation (10) is a formal statement of the law of one price. It is entirely equivalent to equation (4), subject to the conditions required for parity. Equation (10) also suggests an orthogonality condition, which is implied by aggregate profit-maximizing behavior, as represented for a representative agent by the first-order conditions (7) and (8).

Hansen (1982) and Hansen and Singleton (1983) have developed a strategy for estimating the underlying parameters implied by a representative agent's stochastic optimization problem by applying the generalized method of moments (GMM) estimation procedure to orthogonality conditions suggested by the first-order conditions. Following Gallant, we can define a set of parameters, $\tau = \{\beta_0, \beta_1, \beta_2\}$ and denote their true but unknown value as τ_o. We can then define an error function as:

$$e_t(s) = p_t - \beta_0 p_{t+j}^* \beta_1 p_{t+j} \beta_2 \qquad (11)$$

The first-order conditions of the agent's optimization problem then imply:

$$E_t [e_t(s^o)] = 0 \qquad (12)$$

or that the conditional expectation of each of the error functions evaluated at the true parameter value is zero (Tauchen). We can now impose rational expectations, thus requiring the agent's subjective probability distribution to be the same as the distribution governing the random variables contained in the information set, Ω_t. This forms the basis for GMM estimation. Note that we have not required that the information set contain lagged errors. This is an important distinction which will be discussed in detail below. A suitable vector of instruments, z_t (for which $z_t \subset \Omega_t$), allows us to write:

$$m_t(s) = [e_t(s) \otimes z_t] \qquad (13)$$

where $\otimes$ denotes the Kronecker product. By the law of iterated projections, we know:

$$E\ (m_t(\tau^\circ))\ =\ E\ E_t\ [e_t(\tau^\circ)\otimes z_t]\ =\ 0 \tag{14}$$

where E is the unconditional mathematical expectations operator. We can write the first sample moment as:

$$m_n(\tau)\ =\ 1/n\ \Sigma_t = m_t(\tau) \tag{15}$$

and estimate the parameters by forcing $m_n(\tau)$ to $E\ m_n(\tau)$ and solving for τ. The generalized method of moments is a procedure for estimating the parameters $\tau^\circ = \{\beta_0, \beta_1, \beta_2\}$ by attempting to equate the sample moments to the population moments. We express the law of one price in implicit form as:

$$P_t\ -\ \beta_0\ {}_{\{}\pi_{t+j}{}^\beta{}_1 P_{t+j}{}^{*\beta}{}_2\}\ =\ et(\tau) \tag{16}$$

and form the sample moments:

$$m_n(\tau)\ =\ 1/n\ \overset{n}{\underset{t=1}{\Sigma}}\ m_t(y_t,\ x_t,\ \tau) \tag{17}$$

where : $\quad m_t(y_t,\ x_t,\ \tau\ )\ =\ e_t(\tau)\ \otimes z_t \tag{18}$

and equate them to the population moments:

$$m_n(\tau)\ =\ E\ (m_n(\tau^\circ)) \tag{19}$$

Parameter estimates are then obtained by choosing the τ that minimizes the objective function, given by:

$$S(\tau,\ V)\ =\ [n\ m_n(\tau)]'\ V^{-1}\ [n\ m_n(\tau)]\quad \text{with} \tag{20}$$

$$V\ =\ Cov\ \{\ [n\ m_n(\tau)],\ [n\ m_n(\tau)]'\ \} \tag{21}$$

A dynamic correlation structure underlies the error framework of rational expectations problems of this kind. Hansen and Hodrick (1980) have noted that this sort of correlation structure commonly arises in problems where the forecast interval is greater than the sampling interval. Only in cases where the sampling interval equals the forecast interval (i.e., where $j=1$) will the forecast errors be uncorrelated.

Consider again the error function given by equation (11). Because of the delivery lag of j periods, the current value of the error function is not known until j-1 periods have passed. Therefore, $e_{t-1},...,e_{t-j+1}$ are not in the information set Ω_t and thus may be correlated with e_t. Thus, e_t possesses the following properties:

$$E\ [e_t \mid \Omega_t]\ =\ 0 \quad \text{and} \tag{22}$$

$$E\ [e_t\ e_{t-k} \mid \Omega_t]\ \left\{ \begin{array}{ll} \neq\ 0 & \text{for } k+1 \leq j \\[2mm] =\ 0 & \text{for } k+1 > j \end{array} \right. \tag{23}$$

This means that e_t will follow a moving average process of order j-1.

In this analysis, a consistent estimate of the asymptotic covariance matrix which accounts for the serially correlated error structure described above is calculated using a method outlined by Cumby et al. (1983) and Hansen (1982). In addition to accounting for the error correlation, these estimates also allow for conditional heteroskedasticity in the forecast errors. The general estimation strategy involves two steps. In the first stage, preliminary estimates of the forecast errors are calculated from preliminary parameter estimates chosen by (20) with V given as an identity matrix. An estimate of the asymptotic covariance matrix is then calculated by:

$$=\ \sum_{k=-j+1}^{j-1}\ \{\ 1/n\ \sum_{t=1}^{n}\ (u_t \otimes z_t)\ (u_{t-k} \otimes z_{t-k})'\ \} \tag{24}$$

where u_t is the first-stage residual for time t.[1]

3. AN EMPIRICAL APPLICATION TO U.S. AGRICULTURAL COMMODITY PRICES

This section applies the standard model of the law of one price, represented by equation (3), and the expectations-augmented model of the law of one price, represented by equation (4), to an empirical consideration of the law of one price for internationally traded agricultural commodities. If, as has been argued, standard tests of the LOP misrepresent the intertemporal elements of commodity trade and arbitrage, tests of the form given by (4) should find stronger empirical support for the concept of international price parity. The empirical applications are to international markets for U.S. origin wheats and oilseed products. These markets are of special interest because of the prominent role that they play in U.S. agricultural trade. In 1987, the grain and oilseed markets accounted for over 59 percent of the total value of U.S. agricultural exports (U.S. Department of Agriculture, 1988).

The data utilized in this segment of the analysis represent monthly observations of conceptually relevant market data for internationally traded basic homogeneous commodities. Statistical price data for internationally traded commodities are available from a wide range of sources. The oilseed product prices were published by the Foreign Agriculture Service of the USDA in Oilseeds and Products. These prices were quoted monthly at various U.S. interior and export market locations and at Rotterdam. Wheat prices were collected for several varieties of wheat from the International Wheat Council's World Wheat Statistics. These prices are quoted monthly at various U.S. export markets and at Japanese and Rotterdam import markets. The importance of the Rotterdam market derives from its role as a major port of entry and transshipment point for U.S. agricultural products moving into the European market.

Particular characteristics of the data are summarized in an appendix table. However, there are two important issues that should be addressed at this point. First, several of the price series had a small number of missing observations. Because of the importance of the time series structure of the data, deletion of observations was deemed to be too strong a step. As an alternative, portions of the data that were complete or nearly complete were isolated and univariate time series models were used to forecast the missing observations. In no case were there more than two consecutive missing observations or more than five missing data points. The number of missing observations for each series is included in the appendix table. It should be noted that such a procedure may introduce the potential for biases and inconsistencies in the estimation and subsequent statistical inferences.

A second point of significance is that international trade in U.S. agricultural products is customarily invoiced in dollars in both domestic and foreign markets. The result is that prices of U.S. agricultural products in foreign markets are often quoted in dollars. The prices utilized in this section are of this sort. Because trade is conducted solely in dollars, the implied rate of currency exchange is everywhere identically equal to one. These circumstances provide an especially suitable environment for testing the law of one price. The elimination of a direct consideration of exchange rates focuses attention upon the real issue of arbitrage and price behavior. Attempts at separating price and exchange rate effects in empirical tests of the law of one price have been criticized as being too restrictive and without justification by Crouhy-Veyrac et al. (1982). The use of price data quoted in a common currency in international markets represents the "purest" environment possible for testing price parity.

The empirical implementation of the expectations-augmented version of the law of one price as represented by equation (4) requires that a fixed delivery lag be a priori specified. The identification of an appropriate delivery lag for the expectations-augmented model was undertaken through a joint consideration of three important factors. First, prior to any empirical considerations, institutional factors and information of relevance to the markets in question (such as knowledge of trade flows) was utilized to identify the approximate time required to complete an international transaction involving the physical exchange of commodities. Discussions with USDA personnel revealed delivery lags ranging from six to nine weeks for exchange between the U.S. export and Rotterdam import markets. Second, adherence to the LOP for estimators with alternative lag lengths was considered. Finally, the value of the estimation objective function, given by equation (20) was evaluated for alternative estimators. It is important to recognize that each test of the LOP carries with it the augmenting hypothesis that the price relationship (i.e., the delivery lag) be correctly specified. Though every effort has been made to identify accurately the appropriate lag for each market, the choices essentially were made through an informal evaluation. Keeping these conditions in mind, a lag length of $j=2$ months is assumed for the empirical evaluation of the expectations-augmented model.

Under these conditions, the basic tests of the law of one price pursued in this section are of the form:

$$p_{it} = \alpha_0 \, (p_{it}^*)^{\alpha_1} + e_t \tag{25}$$

$$p_{it} - \beta_0 \, E_t\{ \, (p_{it+2}^*)^{\beta_1} \, \} = u_t \tag{26}$$

where e_t and u_t are random disturbances with mean zero. The law of one price is supported for the standard model when the parameter α_1 is found not to be significantly different from one. Similarly, the law of one price is supported for the expectations-augmented model when the parameter β_1 is found not to be significantly different from one.

The standard version of the law of one price as represented by equation (25) initially was estimated using ordinary nonlinear regression techniques. However, the presence of severe autoregressive correlation was detected in the estimates. Such persistence is necessarily precluded by efficient commodity markets. If, as has been argued, a proper consideration of price parity relationships necessarily requires explicit attention to the role of delivery lags, tests of the form of (25) will be misspecified. The presence of such autocorrelation may follow directly from misspecification. The standard model subsequently was reestimated using the conditional nonlinear least squares (NLS) procedure to correct for first-order autoregression. This procedure (discussed in Judge et al. 1985) is asymptotically equivalent to maximum likelihood if the residual errors are normally distributed. Such an explicit correction for autoregressive correlation is often employed in applications of the standard model (e.g., Protopapadakis and Stoll 1983; and Crouhy-Veyrac et al. 1982).

Formal tests of the law of one price for the standard and expectations- augmented versions of the basic model were conducted using the standard F and chi-square representations of the Wald test, respectively. To implement the Wald test, one represents the hypothesis as a functional relationship, given for the parameter set r as:

$$h(r) = 0 \tag{27}$$

We denote h(r), evaluated at $\gamma = \hat{\gamma}$ as h($\hat{\gamma}$). We denote the Jacobian of the estimated function as $\hat{F}$ and the Jacobian of h($\hat{\gamma}$) as $\hat{H}$. The standard F representation of the Wald test statistic is then given by:

$$W = \frac{[\; \hat{h}'(r) \; (\hat{H} \, \hat{C} \, \hat{H}')^{-1} \hat{h}(r) \;]}{q \; s^2} \tag{28}$$

where:

$$C = (\hat{F}'\hat{F})^{-1} \tag{29}$$

Note that s^2 is the mean square error estimate and q represents the number of restrictions being considered in the null hypothesis (the dimension of the h() vector). The chi-square representation of the Wald statistic is given by the numerator of (28). This distinction is necessary in that denominator degrees of freedom are undefined in the moment-based approach of the GMM procedures.

Parameter estimates and hypothesis testing results for the standard version of the law of one price are presented in Table 4.1. Note that ρ represents an autoregressive parameter. In every case except that for the Wheat 5 variable, this autoregressive parameter is found to be highly significant. Also, the price coefficients are highly significant in every case. Overall the data do not appear to be perversely at odds with the LOP theory. The parameter of interest, the price coefficient α_1, lies within the interval [.75, 1.25] in twelve of the fifteen markets.

The formal test of H_0: $\alpha_1 = 1$ is rejected for thirteen of the fifteen cases. The test is supported for sunflowerseed oil and peanut oil. Cottonseed meal, sunflowerseed, and sunflowerseed meal have price coefficients that strongly reject parity. However, the remaining markets have price coefficients that are significantly different from one but are not in strong conflict with the concept of price parity. For wheat, the Rotterdam markets seem to show slightly stronger support for the LOP than the Japanese markets. Soybeans, a commodity of major importance in world agricultural trade, fail to support the standard version of the LOP. When interpreting these results, it is important to recognize the potential for specification biases in estimation and hypothesis testing that may be resident in the standard model. In particular, preceding discussions addressing the fallibility of standard approaches to LOP testing have argued in favor of an alternative model and estimation procedure. The expectations-augmented version of the basic model of the law of one price as represented by equation (26) was estimated using the GMM procedures discussed in the preceding section. The instrumental variables employed in the estimation were:

$$z_{it} = \{1, \; p_{it-1}, \; p_{it-1}^{*}, \; \pi_{it}\} \tag{30}$$

where π_{it} is an exchange rate relevant to trade in the market under consideration[2].

Parameter estimates and hypothesis testing results for the expectations-augmented version of the law of one price are presented in Table 4.2. The price coefficients are highly significant in every case under consideration. The price coefficients for the reformulated version of the LOP are numerically closer to the theoretically implied value of one than are the price coefficients for the standard model in eleven of fifteen cases. In two of the remaining cases, the differences between the estimated price coefficients for the alternative versions of the LOP are less than .008.

The results of formal hypothesis testing for the expectations-augmented version of the LOP support parity in eight of the fifteen cases considered. The LOP is still rejected in every case considering trade between the United States and Japan. However, the results support parity between prices in U.S. and Rotterdam markets in eight

TABLE 4.1. Standard Version of the LOP, Parameter Estimates and Hypothesis Testing Results

Market	α_0	α_1	ρ	P(x>F)	Result
Soybeans (U.S./Rott.)	2.7787 (.6058)[a]	.7943 (.0382)	.8733 (.0495)	.0001	R[b]
Sunflrsd. (U.S./Rott.)	8.4191 (5.2943)	.5647 (.1044)	.9430 (.0478)	.0001	R
Soybean Meal (U.S./Rott.)	.2655 (.0694)	1.2022 (.0450)	.9980 (.0161)	.0001	R
Ctnsd. Meal (U.S./Rott.)	14.2802 (8.9814)	.4875 (.1212)	.7694 (.0667)	.0001	R
Soybean Oil (U.S./Rott.)	.3414 (.1406)	1.1420 (.0609)	.9965 (.0210)	.0217	R
Ctnsd. Oil (U.S./Rott.)	.3475 (.1090)	1.1310 (.0480)	.4988 (.0878)	.0075	R
Sunflrsd. Oil (U.S./Rott.)	.6875 (.3623)	1.0340 (.0777)	.9704 (.0284)	.6626	FTR
Peanut Oil (U.S./Rott.)	.7952 (.6686)	1.0091 (.1182)	.9422 (.0386)	.9388	FTR
Sunflrsd. Ml. (U.S./Rott.)	6.1719 (3.9504)	.5503 (.1248)	.7726 (.0684)	.0005	R

Table 4.1 (continued)

Market	α_0	α_1	ρ	P(x>F)	Result
Wheat 1 (U.S./Japan)	2.0491 (.4168)	.8278 (.0387)	.6187 (.0929)	.0001	R
Wheat 2 (U.S./Japan)	1.9729 (.3900)	.8389 (.0383)	.7779 (.0731)	.0001	R
Wheat 3 (U.S./Japan)	2.5635 (.9230)	.7911 (.0681)	.4578 (.1489)	.0030	R
Wheat 4 (U.S./Japan)	1.8140 (.3669)	.8662 (.0380)	.8357 (.0651)	.0007	R
Wheat 5 (U.S./Rott.)	1.3695 (.2106)	.9260 (.0296)	-.0802 (.0861)	.0136	R
Wheat 6 (U.S./Rott.)	1.3883 (.2038)	.9113 (.0284)	.5072 (.0993)	.0025	R

[a]Numbers in parentheses are asymptotic standard errors.

[b]Hypothesis testing results are at the 5 percent level of significance; FTR - fail to reject, R - reject.

TABLE 4.2.   Expectations Augmented Version of the LOP,
Parameter Estimates and Hypothesis Testing Results

Market	β_0	β_1	$P(x>/^2)$	Result
Soybeans (U.S./Rott.)	.3440 (.1971)[a]	1.1677 (.1035)	.1052	FTR[b]
Sunflrsd. (U.S./Rott.)	.8589 (1.3080)	.9882 (.2700)	.9650	FTR
Soybean Meal (U.S./Rott.)	.5618 (.2338)	1.0908 (.0777)	.2422	FTR
Ctnsd. Meal (U.S./Rott.)	7.6887 (3.7618)	.6131 (.0966)	.0001	R
Soybean Oil (U.S./Rott.)	1.2012 (.7716)	.9656 (.1038)	.7403	FTR
Ctnsd. Oil (U.S./Rott.)	.0649 (.0610)	1.3915 (.1458)	.0072	R
Sflrsd. Oil (U.S./Rott.)	1.2514 (.8618)	.9621 (.1094)	.7291	FTR
Peanut Oil (U.S./Rott.)	4.5991 (3.9193)	.7697 (.1295)	.0753	FTR

Table 4.2 (continued)

Market	β_0	β_1	$P(x>/^2)$	Result
Sflrsd. Meal (U.S./Rott.)	3.0062 (1.4991)	.6898 (.1004)	.0020	R
Wheat 1 (U.S./Japan)	1.5486 (.4126)	.8800 (.0506)	.0178	R
Wheat 2 (U.S./Japan)	2.0527 (.4972)	.8310 (.0468)	.0003	R
Wheat 3 (U.S./Japan)	2.4985 (.8127)	.7990 (.0615)	.0011	R
Wheat 4 (U.S./Rott.)	1.4622 (.3435)	.9025 (.0443)	.0278	R
Wheat 5 (U.S./Rott.)	.7398 (.2690)	1.0476 (.0699)	.4956	FTR
Wheat 6 (U.S./Rott.)	.7114 (.4011)	1.0433 (.1096)	.6927	FTR

[a]Numbers in parentheses are asymptotic standard errors.

[b]Hypothesis testing results are at the 5 percent level of significance; FTR = fail to reject; R = reject.

of eleven cases. With the possible exceptions of cottonseed meal and sunflowerseed meal, none of the commodities are in strong violation of the LOP. It is only for these cases that the price coefficients lie outside the range [.75, 1.25]. It should be noted that, in general, the parameters of the expectations-augmented version of the LOP have larger standard errors than those of the standard version. However, in no case does this result allow the LOP to be supported by the reformulated version and rejected by the standard version when the price coefficient for the reformulated version is actually numerically farther from one. In addition, the standard errors of the standard model are certainly biased if price linkages are not of a contemporaneous nature.

In all, the empirical applications pursued in this segment of the analysis would seem to offer support for contentions that price parity is more likely to be observed when price expectations are given explicit attention. In particular, the LOP is empirically supported in only two of fifteen cases for the standard model but in eight of fifteen cases for the expectations-augmented model. However, empirical support for the LOP remains somewhat limited, suggesting that international price linkages between these markets may not be fully accounted for by the expectations-augmented model.

4. CROSS-CURRENCY TRADE AND INTERNATIONAL FORWARD AND FUTURES MARKETS

The preceding section contained an application of the expectations-augmented model of the law of one price to international markets for U.S. agricultural commodities. The results of this application a rational expectations view of the LOP. However, this segment of the analysis is somewhat specialized in that the application is to a case in which trade and arbitrage activities are transacted in a common currency. In this section, we will extend the basic expectations-augmented model to consider international price relationships for a wider range of commodities. In particular, the analysis will be extended in two important areas. First, the model will be applied to primary commodities that are traded across currencies. This extends the analysis by incorporating the issue of exchange rate passthrough, that is, whether changes in exchange rates are reflected in equilibrating changes in prices for internationally traded commodities. Second, the role of international futures[3] markets will be examined empirically within the context of the basic LOP model. Such markets are relevant to considerations of the temporal elements of trade under uncertainty in that under certain conditions[4] they may offer opportunities for reducing the risks associated with trade and arbitrage activities carried out under conditions of uncertain prices and exchange rates. If such opportunities are present, the issue of price expectations may be secondary to the question of adherence to the law of one price. Instead, the relevant price for comparison becomes the price of a futures contract maturing at a time coincidental with the delivery lag.

The empirical applications pursued in this section utilize monthly average prices quoted in various U.S. and London markets for metals and agricultural commodities over the 1980 through 1987 period. The London prices were collected from selected issues of the London Times. A weekly price was taken as the average of the midweek (Wednesday) afternoon bid and ask quotes. These weekly prices were averaged to obtain a monthly average price. Spot and forward prices for the metal commodities in the London market were quoted at the London Metals Exchange (LME). Futures prices for barley and wheat in London are quoted at the London Grain Futures Market (GAFTA). Soybean meal futures prices in London are quoted at the London Soya Bean Meal Futures Market. Statistical data for monthly observations of U.S. market commodity prices were taken from a variety of sources. Metals prices in various U.S. markets were collected from selected issues of the American Metal Market's Metal Statistics. Barley prices were collected from selected issues of the Commodity Research Bureau's Commodity Yearbook. Monthly futures prices for copper, silver, and aluminum were calculated as the average of daily closing quotes at the Comex Exchange using unpublished data obtained from the Commodity Research Bureau. Data on monthly average spot rates of currency exchange between the United States and the United Kingdom were collected from the International Monetary Fund's International Financial Statistics series. A data series of monthly average forward exchange rates between the United States and the United Kingdom was calculated by averaging weekly quotes taken from the Harris Bank Foreign Exchange Weekly.

Recall that an application of the expectations-augmented version of the LOP requires the identification of appropriate price linkages. These price linkages depend upon the relevant delivery lags as well as on the trading position (i.e., net exporter or importer) of the countries in question. These price linkages were relatively easy to establish in the preceding applications to international markets for U.S.-origin agricultural commodities. However, for the applications pursued in this section, the relevant price linkages are somewhat less obvious. An explicit consideration of salient trade statistics for the metal commodities considered in this section revealed that both the United

States and the United Kingdom were net importers of each commodity over the period under consideration.[5] A delivery lag of j = 3 months was established through a consideration of the minimum sum of squared error criteria for alternative lag structures as well as by the availability of certain futures prices. The three month delivery lag coincides with contract maturity lengths of many of the futures and forward price variables. In particular, for U.K. metals and exchange rates, the futures price variables were constructed by taking the averages of three-month forward prices. For grains, the futures price variable was constructed by taking the average of the two nearest contracts. For metals in the U.S. markets, a futures price variable was constructed by averaging daily closing quotes for the nearest contract. These actions, while attempting to match the implicit price relationships as closely as possible, were often dictated by the availability and quality of data. Hopefully, the effects of any discrepancies resident in the data are minimized by the averaging processes undertaken in data construction.

Under these conditions, the applications of the expectations-augmented model of the LOP to cross-currency exchange in international spot commodity markets are of the form:

$$E_t\{P_{1t+3}\} \quad = \quad \tau_0 \ (E_t\{(P_{1t+3}{}^*)^{\tau}{}_1 \ (\pi_{12t+3})^{\tau}{}_2\}) \qquad (31)$$

Note that we have assumed equal delivery lags between the two importing markets and the corresponding exporting market. In the applications that utilize futures prices and exchange rates, the expected values of the price and exchange rate variables are replaced by analogous futures market variables. Thus, for futures market comparisons between two importing markets, the price relationships are given by:

$$E_t\{{}_tF_{1t+3}\} \quad = \quad \tau_0 \ (E_t\{({}_tF_{1t+3}{}^*)^{\tau}{}_1 \ ({}_t\pi_{12t+3})^{\tau}{}_2\}) \qquad (32)$$

where ${}_tF_{t+3}$ is the futures (forward) price for delivery in time $t+3$ and ${}_t\pi^t_{12t+3}$ is the three-month forward currency exchange rate, both quoted in time t.

The basic expectations-augmented model of the LOP for trade across currencies was applied to a consideration of international price relationships for five metal commodities in U.S. and U.K. spot markets. Again, estimation was accomplished using the GMM procedures, with instrument sets given by the lagged values of each of the variables. Hypothesis testing was accomplished using two alternative test statistics. First, adherence to the LOP was considered for each coefficient (price and exchange rate) individually using the chi square representation of the Wald test statistic. Second, the full hypothesis of adherence to the LOP, given by H_0: $r_1 = r_2 = 1$, was tested using the likelihood ratio test statistic:

$$L = S(\tilde{\theta}, \hat{V}) - S(\hat{\theta}, \hat{V}) \tag{33}$$

where S() represents the optimal value of the sample objective function (sum of squared errors), $\hat{\theta}$ and $\tilde{\theta}$ represent estimates of the unrestricted and restricted parameter sets, respectively, and $\hat{V}$ is the asymptotic covariance matrix estimate from the unrestricted model. These test statistics are discussed in detail in Gallant (1978). The null hypotheses are rejected when W and L exceed the upper α x 100 critical point of the chi-square distribution with q degrees of freedom.

Parameter estimates and hypothesis testing results for the expectations-augmented version of the LOP in international spot metals markets are presented in Table 4.3. The price and exchange rate coefficients are significant in every case. However, the standard errors of the parameter estimates are considerably higher than those obtained in preceding applications of the expectations-augmented model. The parameter estimates do not appear to violate strongly the concept of international price parity. An exception is lead, which has a price coefficient with a value of nearly two. In every case except zinc, the numerical value of the exchange rate coefficient is closer to one than the price coefficient. Also, the exchange rate coefficient has a smaller standard error in every case except zinc. Parameter estimates for application to the silver market show an especially strong adherence to parity. The formal test of full adherence to the LOP is rejected for three of the five cases considered. Similarly, the exchange rate coefficient is significantly different from one in three of five cases. The price coefficients are significantly different from one in two of the five cases. No rejection pattern is apparent in the formal testing, as the rejections are, for the most part, confirmed for each coefficient. The LOP receives especially strong support in the cases of copper and silver. In all, these results can be interpreted as offering somewhat limited support for the expectations-augmented version of the LOP in international metals markets.

The futures market version of the LOP given by equation (31) was applied to a consideration of price parity relationships in international futures markets for three metals and three agricultural commodities.

Parameter estimates and hypothesis testing results are presented in Table 4.4. Several of the coefficients are considerably less significant than has been the case in earlier applications of the model. Barley and wheat appear to be in strong violation of the LOP, each having price and exchange rate coefficients insignificantly different from zero and quite far from one in numerical value. A plausible explanation for this result is that the London markets are for European Community (EC) origin barley and wheat. Such grains may not be of a grade and quality comparable to U.S.-origin grains. It is also quite likely that the EC-origin grain prices reflect the effects of market intervention policies that limit international price transmission. In particular, a variable levy is applied to grain commodities that imported into the EC. Soybean meal is not subject to this levy. In light of the variable levy, an alternative version of the basic model in which the barley and wheat prices were adjusted for the levy was also estimated.[8] The results of the adjusted applications were very similar to those obtained from unadjusted prices. The remaining commodities do not appear to be strongly at odds with the concept of international price parity. The coefficients for silver futures are farther from one in numerical value than those for spot silver, but they are also more significant. Conversely, copper futures appear to support parity to a stronger degree than do prices from the copper spot market in that the coefficients are both closer to one and more significant.

The formal test of full adherence to the LOP in the futures markets is rejected for four of the six cases. Similarly, the exchange rate coefficient is significantly different from one in each of the four cases. The price coefficient is significantly different from one in only one of the six cases. However, this is due in part to the larger standard errors for price coefficient estimates, especially for barley and wheat, and thus does not necessarily indicate failures arising in forward currency exchange markets.

The applications contained in this section expand the basic expectations-augmented model of the LOP in two important areas. First, the model is applied to a consideration of trade across currencies. The results of this application provide limited support for the LOP in international markets for metals. Second, the possible relevance of futures and forward trading was examined within the context of the basic LOP model. The results of these applications provided full support for the LOP in only two of six cases. However, the formal rejection in two of these cases, barley and wheat, may be attributable to heterogeneous commodity comparisons. In all, the results provide mixed support for the LOP in international spot markets and limited support in international futures markets. The

TABLE 4.3. Revised Version of the LOP, Parameter Estimates[a] and Hypothesis Testing[b] Results for Spot Metals Markets

Market	β_1	β_2	T_1	T_2	T_3
Copper (U.S./U.K.)	.7780 (.2220)	.9625 (.0980)	1.00 (.3172)	.15 (.7020)	5.62 (.0602)
Silver (U.S./U.K.)	1.0471 (.1527)	.9984 (.1009)	.10 (.7575)	.01 (.9871)	.19 (.9093)
Tin (U.S./U.K.)	.6578 (.1892)	.7780 (.1055)	3.27 (.0705)	4.42* (.0355)	7.23* (.0269)
Lead (U.S./U.K.)	1.9973 (.3448)	.4998 (.1312)	8.36* (.0038)	14.60 (.0001)	13.80* (.0010)
Zinc (U.S./U.K.)	.8746 (.0953)	.7112 (.1066)	1.73 (.1879)	7.34* (.0067)	10.25* (.0060)

[a]Numbers in parentheses are asymptotic standard errors.

[b]Test Statistics are for the following hypotheses:

$T_1:H_0: \beta_1 = 1,$
$T_2:H_0: \beta_2 = 1,$
$T_3:H_0: \beta_1 = \beta_2 = 1;$

and numbers in parentheses are test probabilities. Note that a "*" indicates rejection of the hypothesis.

TABLE 4.4. Revised Version of the LOP, Parameter Estimates[a]
and Hypothesis Testing[b] Results for Futures
Markets

Market	β_1	β_2	T_1	T_2	T_3
Barley	.6074	.5242	.39	5.49*	8.78*
(U.S./U.K.)	(.6278)	(.2031)	(.5317)	(.0192)	(.0124)
Soybean Meal	1.2058	.8096	2.42	10.10*	31.94*
(U.S./U.K.)	(.1322)	(.0599)	(.1197)	(.0015)	(.0001)
Wheat	.3715	.3308	12.50*	82.72*	89.37*
(U.S./U.K.)	(.1778)	(.0736)	(.0004)	(.0001)	(.0001)
Silver	.9141	.9868	2.73	.06	5.73
(U.S./U.K.)	(.0520)	(.0524)	(.0982)	(.8015)	(.0571)
Aluminum	1.08431	.2472	2.91	11.84*	12.32*
(U.S./U.K.)	(.0495)	(.0718)	(.0882)	(.0006)	(.0021)
Copper	.8752	.9455	1.94	1.42	1.99
(U.S./U.K.)	(.0897)	(.0458)	(.1642)	(.2336)	(.3700)

[a]Numbers in parentheses are asymptotic standard errors.

[b]Test Statistics are for the following hypotheses:

$$T_1: \quad H_0: \beta_1 = 1,$$
$$T_2: \quad H_0: \beta_2 = 1,$$
$$T_3: \quad H_0: \beta_1 = \beta_2 = 1;$$

and numbers in parentheses are test probabilities. Note that
a "*" indicates rejection of the hypothesis.

results may suggest that opportunities for risk reduction in international futures markets are somewhat limited. Finally, it is againimportant to note that the limited support for the LOP contained in these results may suggest that the simple expectations-augmented model may not fully account for the international price linkages between markets.

5. EXTENSIONS TO THE BASIC MODEL OF THE LAW OF ONE PRICE

The basic model of the law of one price which is applied in the preceding sections is dependent upon two simplifying assumptions that may in fact be quite restrictive. The first of these assumptions concerns the manner in which transfer and transaction charges are represented in the basic model. Recall that it was assumed that such costs could be represented as a constant proportion of the domestic nominal price, given by α in equation (5). The second assumption concerns the rate at which future returns from trade and arbitrage activities are discounted. Because of the j-period delivery lag in equation (5), it is necessary to account for foregone interest income. In the expectations-augmented version of the basic model, a constant discounting factor, represented by ß in equation (5), is assumed to represent these interest costs. In this section, the validity of these assumptions will be examined in light of the preceding empirical results which offered limited support for the LOP. An alternative representation of the law of one price that is less restrictive in terms of the empirical representation of interest and transportation charges will also be considered.

The implicit ex-post price relationship between two markets engaged in trade activities that require a two-month delivery lag and that are carried out in a common currency can be written as:

$$p_t = \beta_0 \, (p_{t+2}^*)^\beta{}_1 \qquad (34)$$

Deviations from the LOP can be expressed as values of β_1 that differ from the theoretically implied value of one. The absolute version of the LOP implies a value of one for the multiplicative intercept term β_0. Under the assumption of a constant value of β_0, β_1 can be expressed at each combination of prices as:

$$\beta_{1t} = (\ln p_t \, / \, \ln p_{t+2}^*) - (\ln \beta_0 \, / \, \ln p_{t+2}^*) \qquad (35)$$

For values of β_0 that are reasonably close to one and for large price values, the second term of equation (35) becomes very small. Thus, the second term will be omitted and the price parity relationship will be evaluated at each point for each commodity by considering:

$$\tilde{\beta}_{1it} \; - \; \ln \, p_{it} \, / \, \ln \, p_{it+2}^{*} \qquad (36)$$

Deviations from the absolute version of the expectations-augmented LOP occur for values of $\tilde{\beta}_{1it}$ that are numerically different from one.

The evaluation of deviations from parity conditions will be concentrated upon the oilseed commodities. The price quotes for these commodities are, for the most part, taken over coincidental time periods. This approach focuses attention upon the identification of deviations that occurred across commodity markets in contrast to deviations that were commodity specific. A graphical analysis was undertaken for each of the oilseed commodities. The analyses revealed a high degree of correlation for deviations across commodity markets. Figure 4.1 contains a plot of the deviations from parity conditions, expressed as $1 - \tilde{\beta}_{1it}$. The plot presents the mean value of $1 - \tilde{\beta}_{1it}$ (across all $i = 1...9$ of the oilseed commodities) as well as deviations for soybeans, chosen as a representative commodity. The plot reveals two periods for which significant deviations from parity existed. Large departures from the LOP occurred around June of 1980 and June of 1983. A smaller negative deviation is present in June of 1984.

To undertake an empirical evaluation of the assumptions regarding transactions and transfer charges, it is necessary to have a reasonable measure of these costs. A series of observations on international freight rates for wheat was collected for the period covering July 1975 through December of 1985 from the International Wheat Council's World Wheat Statistics. This series provides a measure of the actual costs of wheat trade between several U.S. export markets (the Gulf, Atlantic, and Pacific ports) and the Rotterdam and Japanese import markets. The freight rates are expressed in dollars per unit terms (an additive representation). An additive charge can be converted to an equivalent proportional rate by dividing through by the domestic nominal price. In the case of proportionally constant transportation charges, such a conversion should produce a series that exhibits little (no) variation.

A proxy measure of the proportional transportation cost parameter for soybeans was calculated by dividing the freight rate series (for wheat trade between the U.S. Gulf ports and Rotterdam) by the U.S. domestic price of soybeans. The resulting series (alpha) is plotted

Figure 4.1

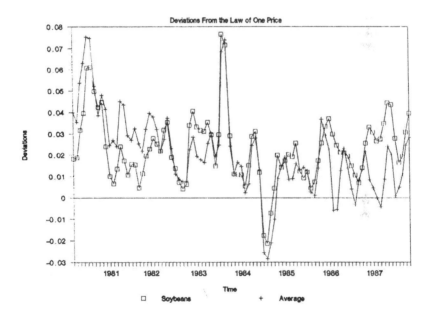

alongside the deviations from parity conditions for soybeans in Figure 4.2. The series displays a considerable degree of variation, ranging froma low of about 2 percent to a high of about 10 percent. The results of this graphical evaluation would seem to question the validity of an assumed constantly proportional representation of transportation charges. The graphical analysis does not appear to reveal a definite pattern of correlation. The transportation cost parameter does show a large shock, which coincides with the first large deviation from parity. The correlation coefficient calculated over the entire series has a value of .3394 (significant at the 5 percent level). The correlation coefficient rises to .5278 (significant at the 5 percent level) when calculated through 1982.

The importance of the link between commodity prices and interest rates has been identified in the context of interest parity conditions (IPC) by Frankel (1984) and Kitchen and Denbaly (1987). The presence of such interest parity conditions for commodities engaged in trade activities is implied by equation (5), where $\beta^j = (1 + r)^{-j}$ (note that r is the one-period rate of interest). Under the conditions of interest parity for commodity prices, changes in the rate of interest will be reflected in equilibrating changes in commodity prices.

An empirical investigation of the assumed constant rate of discounting was undertaken using a series of monthly observations of the interest rate on federal funds collected from the U.S. Department of Commerce's Business Conditions Digest. A two-month discounting factor was calculated as the sixth root of $(1 + r_t)^{-1}$, where r_t is the per annum rate on federal funds. The resulting series (beta) is plotted beside the LOP deviations for soybeans in Figure 4.3. The rate of discounting appears to be quite volatile over the period of study. This graphical representation would seem to suggest that the assumption of a constant discounting rate is fundamentally invalid. Figure 4.3 also reproduces the plot of deviations from parity conditions for soybeans. Correlation between the two series appears to be somewhat limited. However, the large deviation from parity that occurred in June of 1980 appears to coincide with a large positive shock to the discounting factor, brought about by a policy shift implemented by the Federal Reserve Board in early 1980. The apparent correlation continues through 1982. It would appear difficult to conclude that the series are correlated over the post-1982 period. The bivariate correlation coefficient for the entire series is .1909 (significant at the 5 percent level). However, for the period spanning January 1980 through December of 1982, the series have a correlation coefficient of .7803 (significant at the five percent level). In all, the graphical analyses would seem to question the validity of assumptions regarding the

Figure 4.2

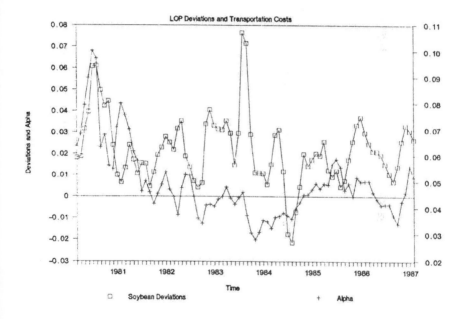

Figure 4.3

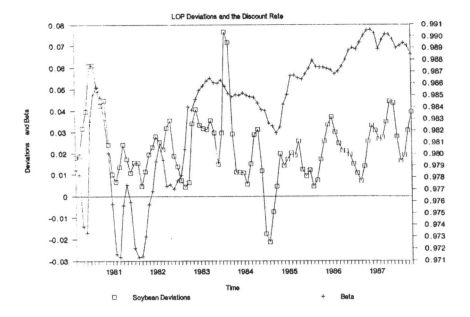

stability of interest and proportional transportation charges. Proxy representations of these charges suggest a high degree of volatility inboth series. However, the graphical analyses do not provide any definite conclusions regarding the source of deviations from parity conditions in the oilseed markets. The figures reveal two periods of relatively large departures from parity. The first of these periods appears to coincide with large shocks to both interest rates and transportation charges. However, the second large deviation does not appear to coincide with shocks to either of these series. In light of these results, it is of interest to consider an alternative model that is less restrictive in terms of the representation of transportation and foregone interest costs.

Consider again the expectations-augmented version of the law of one price. With transportation costs represented in the per-unit additive form, this relationship can be expressed as:

$$P_t = (\beta_t^j)(E_t P_{t+j}^*) - T_t \qquad (37)$$

where $\beta_t = (1 + r_t)^{-1}$ is the (variable) rate of discounting and T_t is the per-unit transportation charge. Because freight rates are unavailable for all of the commodities under consideration, it may be possible to proxy the transportation charges in an empirical application by replacing T_t with:

$$\hat{T}_t = f(TW_t) \qquad (38)$$

where TW_t is the observable freight rate for wheat. It would be expected that transportation charges are highly correlated across all basic agricultural commodities. Thus, in the applications to follow, $f()$ is taken to be a linear function. This representation was chosen from a consideration of alternative functional forms. Transportation costs are therefore represented by:

$$\hat{T}_t = r_0 + r_1 TW_t \qquad (39)$$

It is important to recognize that such a representation may misrepresent actual transportation and transaction costs to some degree.

Under these conditions, the expectations-augmented version of the basic model of the law of one price can be expressed as:

$$P_t \;=\; E_t \,\{\Theta_0 \;+\; B_t{}^J (P_{t+j}{}^*)\Theta_2 \;-\; \Theta_3 TW_t \,\} \tag{40}$$

where $\theta = \{\theta_0, \theta_2, \theta_3\}$ is the parameter set to be estimated. Adherence to the law of one price (subject to the interest parity conditions and the proxy measure of transportation costs) is confirmed for values of Θ_2 that are not significantly different from one. It is important to recognize that such a specification includes the augmenting hypothesis of adherence to the conditions of interest parity for an internationally traded commodity. Thus, it is of interest to examine the interest parity conditions explicitly prior to the estimation of the revised model. A series of monthly observations of the two-month rate of discounting $(B_t{}^2)$ was constructed from the aforementioned interest rate on federal funds. To examine the interest parity conditions formally, the following equation was applied to each of the grain and oilseed markets:

$$P_t \;=\; E_t \,\{\Theta_0 \;+\; (B_t{}^J)\Theta_1 \; (P_{t+j}{}^*)\Theta_2 \;-\; \Theta_3 TW_t \,\} \tag{41}$$

Adherence to the conditions of interest parity (subject to the proxy measure of transportation charges) is verified for values of θ_1 that are not significantly different from one. Equation (41) was estimated for each of the grain and oilseed commodities using the generalized method of moments procedures. The freight rate for wheat trade between the U.S. Gulf ports and Rotterdam was utilized for TW_t for all of the oilseed commodities. The rate relevant to trade between the markets in question was utilized for the wheat commodities. The instrumental variables employed in the estimation were:

$$z_{it} \;=\; \{1, \; P_{it-1}, \; P_{it-1}{}^*, \; TW_{it-1}, \; r_{t-1}, \; \pi_{it}\} \tag{42}$$

The hypothesis of (uncovered) interest parity for traded commodities was formally tested using the chi-square representation of the Wald test statistic. The results of formal hypothesis testing[7] are summarized in Table 4.5. The results formally support the conditions of interest parity in thirteen of the fifteen cases under consideration. However, this apparent support may be attributable to large standard errors on the estimates of θ_1. The apparent lack of complete support for the interest parity conditions is not surprising in light of the inconclusive evidence regarding correlation between LOP deviations and the discounting factor. The results are very similar to those of Kitchen and Denbaly (1987), who attribute the lack of complete support to a large variance of the expectation errors. The lack of complete support for the conditions of interest parity might be attributable to improper proxy measures of transportation charges or to a high degree of correlation between the calculated discounting factor and the transportation charges. Such correlation may complicate the estimation of individual parameters.

In light of the results regarding adherence to the interest parity conditions and the limitations of the proposed revisions to the basic expectations-augmented model, it is important to view the results to follow with an appropriate amount of caution. Under the conditions outlined above, the revised model of the law of one price, represented by equation (40), was applied to each of the wheat and oilseed commodities. Recall that this specification carries the interest parity conditions as a maintained hypothesis. Adherence to the law of one price is verified for values of θ^2 that are not significantly different from one. Estimation of equation (40) was carried out using the generalized method of moments procedures and the instrument set given by (42). Again, the formal test of adherence to the LOP was conducted using the chi-square representation of the Wald test statistic.

Parameter estimates and hypothesis testing results for the revised version of the basic model are presented in Table 4.6. The price coefficients have very small standard errors, indicating a high degree of statistical significance. The price coefficients are quite close to one in every case, falling in the range [.97, 1.03] in eleven of the fifteen cases. The largest deviation from parity occurs with sunflowerseed meal, which has a price coefficient of .9148. Results for the transportation cost parameter, θ_3, are somewhat less encouraging. This coefficient is significant at the 5 percent level in seven of the fifteen cases. If the freight rates for wheat were a perfect representation of the true costs associated with commodity trade, one would expect the values of the transportation cost parameters to be reasonably close to one. As would be expected, the transportation cost parameters are relatively close to one for the wheat commodities (for which the rates apply directly). An exception is Wheat 6, which has a negatively valued (but insignificant) transportation cost parameter. For the remaining commodities, the freight rates for wheat would appear to be an imperfect (but not necessarily invalid) representation of the costs associated with commodity exchange.

The results of formal hypothesis testing for the revised version of the basic LOP model support parity in thirteen of the fifteen cases under consideration. However, it should be noted that very small standard errors allow formal rejection of the LOP for sunflowerseed meal and Wheat 5, which have price coefficients that are reasonably close to one in numerical value. In particular, the price coefficient for

TABLE 4.5. Uncovered Interest Parity Conditions, Hypothesis
Testing Results

Commodity (Markets)	Approximate $Pr(x>/^2)$	Result
Soybeans (U.S./Rotterdam)	.0753	FTR[a]
Sunflowerseed (U.S./Rotterdam)	.3321	FTR
Soybean Meal (U.S./Rotterdam)	.0351	R
Cottonseed Meal (U.S./Rotterdam)	.3210	FTR
Soybean Oil (U.S./Rotterdam)	.4020	FTR
Cottonseed Oil (U.S./Rotterdam)	.7525	FTR
Sunflowerseed Oil (U.S./Rotterdam)	.8153	FTR
Peanut Oil (U.S./Rotterdam)	.2681	FTR
Sunflowerseed Meal (U.S./Rotterdam)	.6657	FTR

Table 4.5 (continued)

Commodity (Markets)	Approximate $Pr(x>/^2)$	Result
Wheat 1 (U.S./Japan)	.0017	R
Wheat 2 (U.S./Japan)	.5518	FTR
Wheat 3 (U.S./Japan)	.2123	FTR
Wheat 4 (U.S./Japan)	.9926	FTR
Wheat 5 (U.S./Rotterdam)	.4785	FTR
Wheat 6 (U.S./Rotterdam)	.0787	FTR

[a]Results are at the 5 percent level of significance.

TABLE 4.6. Revised Version of the LOP, Parameter
Estimates and Hypothesis Testing Results

Market	θ_0	θ_2	θ_3	$P(x>/^2)$	Result
Soybeans	-53.863	1.0253	.8963	.2855	FTR[b]
(U.S./Rott.)	(29.5248)[a]	(.0237)	(1.7447)		
Sunflrsd.	-47.1555	1.0266	5.1309	.5067	FTR
(U.S./Rott.)	(57.0880)	(.0401)	(2.6349)		
Soybean Meal	-20.1301	1.0204	1.8848	.2196	FTR
(U.S./Rott.)	(18.6920)	(.0166)	(1.0732)		
Ctnsd. Meal	76.0320	.9943	5.8129	.7332	FTR
(U.S./Rott.)	(13.2098)	(.0167)	(1.1792)		
Soybean Oil	48.9293	.9928	3.6163	.6992	FTR
(U.S./Rott.)	(46.1870)	(.0187)	(3.2702)		
Ctnsd. Oil	-194.8320	1.0080	-4.8353	.6453	FTR
(U.S./Rott.)	(66.8662)	(.0174)	(2.7385)		
Sflrsd. Oil	43.8963	1.0260	12.2568	.0862	FTR
(U.S./Rott.)	(47.8982)	(.0151)	(2.9700)		
Peanut Oil	293.4883	1.0129	34.8598	.4671	FTR
(U.S./Rott.)	(103.9876)	(.0178)	(9.2810)		
Sflrsd. Meal	26.1290	.9140	2.5258	.0001	R
(U.S./Rott.)	(8.1073)	(.0203)	(.7413)		

Table 4.6 (continued)

Market	h_0	h_2	h_3	$P(x>/^2)$	Result
Wheat 1 (U.S./Japan)	-14.7749 (10.0180)	1.0309 (.0195)	1.6763 (.5271)	.1127	FTR
Wheat 2 (U.S./Japan)	3.7924 (7.0983)	.9977 (.0157)	.9387 (.3606)	.8812	FTR
Wheat 3 (U.S./Rott.)	-5.9607 (13.7183)	1.0257 (.0250)	1.7588 (.6938)	.3039	FTR
Wheat 4 (U.S./Japan)	-2.3699 (8.5510)	1.0185 (.0156)	1.6793 (.5259)	.2352	FTR
Wheat 5 (U.S./Rott.)	-33.3982 (12.8970)	1.0493 (.0156)	2.3758 (.7825)	.0016	R
Wheat 6 (U.S./Rott.)	7.4813 (17.5005)	.9529 (.0373)	-1.0218 (.7799)	.2067	FTR

[a]Numbers in parentheses are asymptotic standard errors.

[b]Hypothesis testing results are at the 5 percent level of significance; FTR = fail to reject; R = reject.

Wheat 5 has a value of 1.093 but still formally rejects the hypothesis of price parity. It is important to interpret these results for theireconomic significance as well asfor their statistical significance. In aneconomic sense, these results strongly affirm adherence to the LOP. Conclusions drawn from a statistical rejection of the LOP in the presence of such small standard errors are of diminished importance to the economic interpretation of the empirical results.

In all, the revised version of the law of one price would appear to offer very strong support for the concept of price parity in international markets for primary homogeneous commodities. In particular, the inclusion of a variable rate of discounting and variable transportation costs appears to offer significant improvements in the empirical performances of the basic expectations-augmented model[8]. However, it should again be noted that the revised version of the LOP developed and applied in this section imposes a particular form of the interest parity conditions on the international price relationships. These conditions, while theoretically valid, are not given complete empirical support in independent hypothesis testing. Thus, these results should be interpreted with caution in that they may be subject to specification biases.

6. CONCLUDING REMARKS

This paper has addressed the issue of price behavior in international markets for several important primary commodities. In particular, adherence to the law of one price is evaluated through theoretical and empirical considerations. The LOP has received a great deal of attention in recent economic applications. In general, the empirical evidence has concluded in favor of rejecting adherence to the LOP in international commodity markets. This analysis has argued that such rejections may have resulted in part from a failure to consider the intertemporal aspects of international commodity arbitrage.

Overall, the paper produces some tentative conclusions regarding adherence to the LOP in international commodity markets. In particular, the presence of delivery lags implies an important role for price and exchange rate expectations. Applications of models recognizing the role of expectations provide stronger empirical support for the LOP than is commonly obtained in standard analyses. Thus, a preliminary conclusion of the study is that international commodity price linkages depend in an essential way upon agents' expectations of future prices. The implication is that adherence to the LOP is much more likely to be observed for expected prices[9].

The results of this paper may have important implications for the appropriate approach to theoretical and empirical modeling of international trade. The standard approach in applied trade modeling is explicitly to assume adherence to the LOP for contemporaneous prices in absolute terms. However, the empirical evidence would seem to question the validity of such suppositions. In contrast, the results of this study suggest that the importance of delivery lags and price expectations should be explicitly recognized in any analysis of international trade in basic homogeneous commodities. This is of particular importance to empirical analyses utilizing quarterly or monthly data for which large deviations from parity conditions are common.

Although this investigation provides some preliminary answers to questions of adherence to the LOP in international commodity markets, several important questions remain open to consideration. Of greatest importance is the fact that adherence to the LOP remains incomplete in even the best of the empirical results of the study. This finding may be attributable to the quality of price data utilized in the empirical analyses. In particular, although every effort was made to guarantee homogeneous commodity comparisons, the price data may still reflect the influences of residual product differentiation. It is also likely that the time-wise aggregation utilized to construct monthly price observations misrepresents actual arbitrage behavior to some degree. This result may also suggest price linkages that are more complex that those considered here. Finally, the representation of transportation and transactions costs may fail to recognize many of the costs associated with international commodity trade and arbitrage. Future research may benefit from considering additional factors and alternative models that may provide stronger support for the LOP.

APPENDIX TABLE 1. Description of Price Variables

Commodity	Description	n	n-miss	period
U.S. Soybeans	U.S. farm level average	105	0	1/80-9/88
Rotterdam Soybeans	No. 2 U.S. origin, cif	105	0	1/80-9/88
U.S. Sunflowerseed	U.S. farm level average	89	1	1/80-5/87
Rotterdam Sunflowerseed	U.S./Canadian origin, cif	89	4	1/80-5/87
U.S. Soybean Meal	Decatur, 44% protein, fob	105	4	1/80-9/88
Rotterdam Soybean Meal	U.S. origin, 44% protein, cif	105	0	1/80-9/88
U.S. Cottonseed Meal	Memphis, 41% protein, fob	105	0	1/80-9/88
Rotterdam Cottonseed Meal	Denmark, 38% protein, cif	105	1	1/80-9/88
U.S. Soybean Oil	Decatur, crude average	105	0	1/80-9/88
Rotterdam Soybean Oil	Dutch ex-mill, fob	105	0	1/80-9/88
U.S. Cottonseed Oil	Valley Points, fob	105	1	1/80-9/88
Rotterdam Cottonseed Oil	U.S. origin, cif	105	2	1/80-9/88
U.S. Sunflowerseed Oil	Minneapolis, fob	105	0	1/80-9/88
Rotterdam Sunflowerseed Oil	Rotterdam ex-mill	105	0	1/80-9/88
U.S. Peanut Oil	Southeastern mills, fob	105	0	1/80-9/88
Rotterdam Peanut Oil	Rotterdam, any origin, cif	105	0	1/80-9/88
U.S. Sunflowerseed Meal	Minneapolis 41% protein, fob	105	0	1/80-9/88
Rotterdam Sunflowerseed Meal	Argentina/Urguay origin 38% protein, cif	105	0	1/80-9/88
U.S. Wheat 1	No. 2 Hard Winter 13%, Gulf fob	72	0	11/75-10/81

Appendix Table 1 (continued)

Commodity	Description	n	n-miss	period
Japan Wheat 1	U.S. origin No. 2 Hard Winter 13%, cif	72	5	11/75-10/81
U.S. Wheat 2	No. 2 Western White 13%, Pacific fob	79	0	11/75-10/81
Japan Wheat 2	U.S. No. 2 Western White 13%, cif	79	2	7/75-1/82
U.S. Wheat 3	No. 2 Dark Northern Spring 14%, Gulf fob	79	0	7/75-1/82
Japan Wheat 3	U.S. No. 2 Dark North Spring 14%, cif	79	5	7/75-1/82
U.S. Wheat 4	No. 2 Dark North Spring 14%, Pacific fob	79	0	7/75-1/82
Japan Wheat 4	U.S. No. 2 Dark North Spring 14%, cif	79	5	7/75-1/82
U.S. Wheat 5	No. 2 Dark North Spring 14%, Gulf fob	138	0	7/75-1/87
Rotterdam Wheat 5	U.S. No. 2 Dark North Spring 14%, cif	138	0	7/75-1/87
U.S. Wheat 6	No. 2 Soft Red Winter, Atlantic fob	79	0	7/79-2/86
Rotterdam Wheat 6	U.S. No. 2 Soft Red Winter, cif	79	0	7/79-2/86

Appendix Table 1 (continued)

Commodity	Description	n	n-miss	period
U.S. Barley (Spot)	No. 2 or Better, Minneapolis	73	0	1/80-8/86
London Barley (Futures)	E.C. Origin, London Gr. Fut. Market	73	1	1/80-8/86
U.S. Soybean Meal (Spot)	Decatur 44% protein, fob	42	0	1/80-6/83
London Soybean Meal (Fut.)	London Soya Bean Futures Market	42	0	1/80-6/83
U.S. Wheat (Spot)	No. 2 Dark N. Spring, Gulf fob	74	0	1/80-2/86
London Wheat (Futures)	E.C. Origin Wheat, London Gr. Fut. Market	74	1	1/80-2/86
U.S. Copper (Spot)	Spot price, Comex	72	0	1/80-12/85
London Copper (Spot)	Wire Bars, London Metals Exc.	72	0	1/80-12/85
U.S. Silver (Spot)	Handy and Harman, NY .999 fine	96	0	1/80-12/87
London Silver (Spot)	Spot price, London Metals Exc.	96	0	1/80-12/87
U.S. Tin (Spot)	Straits Tin, NY Alloyer price	70	0	1/80-10/85
London Tin (Spot)	Standard Tin, London Metals Exchange	70	0	1/80-10/85
U.S. Lead (Spot)	Primary producers, NY common	96	0	1/80-12/87
London Lead (Spot)	Spot price, London Metals Exc.	96	0	1/80-12/87
U.S. Zinc (Spot)	Prime Western Slab, U.S. del. basis	84	0	1/80-12/86
London Zinc (Spot)	Standard Zinc, London Metals Exchange	84	0	1/80-12/86

Appendix Table 1 (continued)

Commodity	Description	n	n-miss	period
U.S. Aluminum (Futures)	Comex Futures price	49	0	12/83-12/87
London Aluminum (Forward)	3-Month Forward, London Metals Exchange	49	0	12/83-12/87
U.S. Silver (Futures)	Comex Futures price	96	0	1/80-12/87
London Silver (Forward)	3-Month Forward, London Metals Exchange	96	0	1/80-12/87
U.S. Copper (Futures)	Comex Futures price	62	0	1/80-2/85
London Copper (Forward)	3-Month Forward, London Metals Exchange	62	0	1/80-2/85

NOTES

1. It should be noted that for j > 1, equation (24) is not guaranteed to be positive definite by construction. Because obtaining parameter estimates requires that we minimize the objective function $S(\tau, V)$, given by (20), the matrix must be positive definite. Cumby et al. (1983) note that is the sum of the autocovariance matrices of the vector process $\{e_t \, P \, z_t\}$ and is therefore equal to the spectral density matrix of this process evaluated at frequency zero. From this fact, appeal can be made to alternative spectral estimators which are positive definite by construction (e.g., a Parzen weighted sum). However, as Hansen and Singleton (1982) note, these estimators fail to exploit the fact that only a finite number of lags (i.e., k = j-1) should be used in constructing. For the empirical applications that follow, the estimated asymptotic covariance matrix was positive definite in every case and the empirical applications were thus able to utilize the correct order of unweighted lag.

2. Specifically, π_{it} is a composite index of U.S./European Community exchange rates (E.C.U.s) for the oilseed applications; the U.S./Japanese exchange rate for the applications considering wheat trade between the United States and Japan; and the U.S./Netherlands exchange rate for the remaining applications.

3. While important distinctions exist between forward and futures markets, the two are treated as being identical for the purposes of this analysis. Thus, wherever the term "futures market" appears, one may also read "forward market."

4. In particular, agents may be able to eliminate the risks associated with intertemporal commodity exchange through actions undertaken in futures and forward markets if the following conditions are satisfied: a forward or futures market for the commodity under trade exists in the importing market; the contracts are of a grade and quality identical to that of the commodity under trade; the futures prices are free from any risk premium or backwardation effects that might inhibit their usefulness as a risk-reducing instrument; the points of delivery specified in the futures contract identically match the delivery point relevant to trade; and contracts that match the delivery lags are available.

5. Trade statistics for the six metals under consideration were collected from the <u>Bureau of Mines Minerals Yearbook</u> and are summarized in Goodwin (1988).

6. Data series on the monthly average variable levies for barley and wheat were assembled from data collected from the Commission of the European Community's <u>Landbrugmarkeder Agrarmakte</u> (Agricultural Markets). The levies were removed from the London prices in a revised application of the model. These applications failed to improve the results of the basic model. Thus, the failure of the basic model is likely attributable to heterogeneity between U.S.- and E.C.-origin grains.

7. Unconstrained parameter estimates are not presented here. However, it should be acknowledged that these estimates were often numerically quite far from one and often had relatively large standard errors. The parameter estimates were very similar to those obtained by Kitchen and Denbaly (1987).

8. An underlying goal of this revised analysis was to identify and incorporate conditions that may have contributed to deviations from parity conditions. To this end, an alternative form of equation (40) which included indicator variables for the two periods of significant deviations was also estimated. The indicator variables were found to be insignificant in the revised model.

9. An alternative application of the expectations-augmented model of the LOP was also considered for weekly internal agricultural prices in different member countries of the European Community (EC). A priori expectations were that prices between member countries were closely linked because of common agricultural policies. However, the results indicated almost no discernable relationship between weekly prices across the European states' borders. Such a finding provides a partial explanation for the existence of monetary compensation amounts (MCAs), which are utilized to equalize prices in response to exchange rate changes.

REFERENCES

Cassel, G. (1918). "Abnormal Deviations in International Exchanges." Economic Journal 28: 413-15.

Chambers, R. G. and R. E. Just (1979). "A Critique of Exchange Rate Treatment in Agricultural Trade Models." American Journal of Agricultural Economics 61: 249-57.

Crouhy-Veyrac, L., M. Crouhy, and J. Melitz (1982). "More About the Law of One Price." European Economic Review 18: 325-44.

Cumby, R., J. Huizinga, and M. Obstfeld (1983). "The Two-Step Two-Stage Estimator." Journal of Econometrics 21: 333-55.

Dornbusch, R. (1976). "The Theory of Flexible Exchange Rate Regimes and Macroeconomic Policy." Scandinavian Journal of Economics 78: 255-75.

Frankel, J. A. (1984). "Commodity Prices and Money: Lessons From International Finance." American Journal of Agricultural Economics 66: 560-66.

Frenkel, J. (1981). "Flexible Exchange Rates, Prices, and the Role of News: Lessons From the 1970s." Journal of Political Economy 89: 665-706.

_____ (1978). The Economics of Exchange Rates. Reading, Massachusetts: Addison-Wesley Co.

Frenkel, J. and H. Johnson (1976). The Monetary Approach to the Balance of Payments. Toronto: University of Toronto Press.

Gailliot, H. J. (1970). "Purchasing Power Parity as an Explanation of Long Term Changes in Exchange Rates." Journal of Money, Credit, and Banking 2: 348-57.

Gallant, A. R. (1987). Nonlinear Statistical Models. New York: Wiley.

Genberg, Hans (1978). "Purchasing Power Parity Under Fixed and Flexible Exchange Rates." Journal of International Economics 8 (May): 247-76.

Goodwin, Barry K. (1988). "Empirically Testing the Law of One Price in International Commodity Markets: A Rational Expectations Approach." Ph.D. Thesis, Department of Economics and Business, North Carolina State University, Raleigh.

Hansen, L. P. (1982). "Large Sample Properties of Generalized Method of Moments Estimators." Econometrica 50: 1029-55.

Hansen, L. P. and R. J. Hodrick (1980). "Forward Exchange Rates as Optimal Predictors of Future Spot Rates: An Econometric Analysis." Journal of Political Economy 88: 829-53.

Hansen, L. P. and K. J. Singleton (1982). "Generalized Instrumental Variables Estimators of Nonlinear Rational Expectations Models." Econometrica 50: 1269-86.

_____ (1983). "Stochastic Consumption, Risk Aversion, and the Temporal Behavior of Asset Returns." Journal of Political Economy 91: 249-65.

Herlihy, Michael T. (1988). Agriculture and Trade Policy Branch, United States Department of Agriculture, personal communication with the author (February).

Hodgson, J. S. and P. Phelps (1975). "The Distributed Impact of Price Level Variation on Floating Exchange Rates." Review of Economics and Statistics 57: 58-64.

International Monetary Fund (Various Issues). International Financial Statistics, Washington, D.C.

International Wheat Council (Various Issues). World Wheat Statistics, London.

Isard, P. (1977). "How Far Can We Push the Law of One Price?" American Economic Review 67: 942-49.

Jabara, C. L. and N. E. Schwartz (1987). "Flexible Exchange Rates and Commodity Price Changes: The Case of Japan." American Journal of Agricultural Economics 69: 580-90.

Jain, A. K. (1980). Commodity Futures Markets and the Law of One Price. Michigan International Business Studies Number 16, Graduate School of Business Administration, Ann Arbor: University of Michigan.

Judge, G. G., W. E. Griffiths, R. C. Hill, H. Lutkepohl, and T. Lee (1985). The Theory and Practice of Econometrics. Second Edition. New York: Wiley.

Kitchen, J. and M. Denbaly (1987). "Arbitrage Conditions, Interest Rates, and Commodity Prices." Journal of Agricultural Economic Research 39: 3-11.

Kravis, I. and R. Lipsey (1978). "Price Behavior in Light of Balance of Payment Theories." Journal of International Economics 8: 193-246.

Marshall, A. (1926). "Memoranda and Evidence Before the Gold and Silver Commission (1888)." in Official Papers. London: Macmillian, pp. 17-195.

Mill, J. S. (1929). Principles of Political Economy. London: Longmans and Green.

Niehans, J. (1984). International Monetary Economics. Baltimore: Johns Hopkins University Press.

Officer, L. (1982). Purchasing Power Parity and Exchange Rates: Theory, Evidence, and Relevance. Greenwich, CT: JAI Press.

_____ (1986). "The Law of One Price Cannot Be Rejected: Two Tests Based on the Tradable/Nontradable Goods Dichotomy." Journal of Macroeconomics 8: 159-82.

_____ (1988). "The Law of One Price: Two Levels of Aggregation." Paper presented at the Winter Meeting of the International Agricultural Trade Research Consortium, San Antonio (December).

Protopapadakis, A. and H. Stoll (1983). "Spot and Futures Prices and the Law of One Price." Journal of Finance 38: 1431-56.

_____ (1986). "The Law of One Price in International Commodity Markets: A Reformulation and Some Formal Tests." Journal of International Money and Finance 5: 335-60.

Richardson, D. J. (1978). "Some Empirical Evidence on Commodity Arbitrage and the Law of One Price." Journal of International Economics 8: 341-51.

Rogalski, R. J. and J. D. Vinso (1977). "Price Level Variations as Predictors of Flexible Exchange Rates." Journal of International Business Studies (Summer): 71-81.

Rosen, S. (1974). "Hedonic Prices and Implicit Markets." Journal of Political Economy 82: 34-55.

Tauchen, G. (1986). "Statistical Properties of Generalized Method of Moments Estimators of Structural Parameters Obtained from Financial Market Data." Journal of Business and Economic Statistics 4: 397-416.

U.S. Department of Agriculture, Economic Research Service (1988). Foreign Agricultural Trade of the United States, 1987 Supplement. Washington, D.C.: USGPO.

U.S. Department of Agriculture, Foreign Agriculture Service (Selected dates). Grain Market Outlook. Washington, D.C.: USGPO.

U.S. Department of Agriculture, Foreign Agriculture Service (Selected issues). Oilseeds and Products. Washington, D.C.: USGPO.

U.S. Department of Commerce (Selected dates). Business Conditions Digest. Washington, D.C.: USGPO.

Wihlborg, C. (1979). "Flexible Exchange Rates, Currency Risks, and the Integration of Capital Markets." in Inflation in Open Economics. Amsterdam: North-Holland.

5

Exchange Rates, Interest Rates, and Agriculture: A Macroeconomic View from Down Under[1]

L. Paul O'Mara

1. INTRODUCTION

Macroeconomic influences on agriculture, particularly exchange rate influences, were a key focus of the August 1986 meeting of the International Agricultural Trade Research Consortium. In many of the papers, the emphasis was on the U.S. experience and on the results of some associated empirical and theoretical research [Paarlberg and Chambers (1988)]. The objective of the present paper is to complement that earlier material by reviewing the recent Australian experience and economic research in the area of exchange rates, interest rates and agriculture.

Australia adopted something approaching a floating exchange rate only in 1983, some years later than many other countries, although a degree of flexibility along the lines of a "crawling peg" had been introduced in 1976. However, it had long been recognized by Australian economists that a fixed nominal exchange would neither prevent nor eliminate the need for movements in the real exchange rate from time to time [Swan (1955, 1960); Salter (1959)].

The underlying causes of longer-term movements in real exchange rates and the implications of such movements for the Australian farm sector received considerable attention after the mid-1970s [for example, Gregory (1976); Snape (1977); Stoeckel (1979); O'Mara, Carland and Campbell (1980)]. Following the floating of the Australian dollar in 1983, major swings in the nominal and real exchange rate and in interest rates stimulated a new round of research in these areas. Because the real exchange rate and real interest rates are probably the two domestic macroeconomic variables with the greatest potential to affect the Australian farm sector, they have also been high on the list of research priorities in the Australian Bureau of Agricultural and Resource Economics (ABARE). In what follows, the focus is on the forces driving exchange rates and interest rates and the outlook for those variables rather than on the detailed microeconomic effects of movements in these variables on the farm sector. The latter issues have been addressed at some length by, for example, Martin and Shaw (1986) and Higgs (1986).

In the next section, recent macroeconomic developments in Australia are reviewed briefly, particularly with respect to exchange rates and interest rates and the associated issues of the current account balance, the terms of trade and foreign debt levels. The theoretical models that have underpinned and provided the intuitive basis for much of the recent Australian research are outlined in section 3. Some of the key components of that research are drawn together and discussed in section 4, with an emphasis on the ABARE contributions. Taken together, that research has permitted ABARE to place a particular interpretation on the recent exchange rate and interest rate developments in Australia, and on that basis to make medium-term projections of those variables. In section 5, some recent ABARE research on international current account imbalances and exchange rate movements is discussed briefly. Finally, section 6 contains some concluding remarks, with particular reference to the sustainability of present key exchange rates.

2. RECENT ECONOMIC DEVELOPMENTS IN AUSTRALIA

A crucial feature of Australia's economic performance during the 1980s has been a marked increase in the deficit in the current account. Since the early 1960s, the current account deficit had averaged around 2 percent of GDP, but during the 1980s the average has been around 5 percent, peaking at about 6 percent in 1985-86 (see Figure 5.1). Since then a significant improvement has occurred, with the deficit in 1988-89 expected to be around 3 percent of GDP.

These developments on the current account have mirrored, to some extent, a sharp decline in Australia's terms of trade during the mid-1980s (Figure 5.2), followed by a substantial recovery more recently. It is clear from Figure 5.2, however, that Australia's terms of trade have long been subject to major volatility, and the experience of recent years has not really been atypical.

Australia's real exchange rate--defined as the nominal exchange rate against major trading partners weighted by general trade shares and adjusted for inflation differentials--fell (devalued) dramatically, by around 37 percent, between the December quarter 1984 and the September quarter 1986 (Figure 5.3). Though estimates of the real exchange rate over long periods of history are necessarily uncertain, there is some evidence that the fall in the real exchange rate in the mid-1980s was one of the sharpest this century and carried it to the lowest level this century [McKenzie (1986)]. Since the latter part of

1986, the real exchange rate has risen markedly, although by the September quarter 1988 it was still around 20 percent below its 1984 level.

The deterioration in the current account balance and the overall decline in the real exchange rate since 1984 have both contributed to a rapid rise in Australia's net foreign debt, from less than 10 percent of GDP in the early 1980s to around 30 percent in 1988 (Figure 5.4). The devaluation contributed to the rise because much of Australia's foreign debt is contracted in foreign currency. Stabilizing and then gradually reducing the debt ratio has emerged as a key priority for Australian policy makers because the level and rate of growth of this ratio is widely seen as a major influence on sentiment in financial markets and therefore on the presence and size of a risk premium in the Australian interest rate structure.

The sharp fall in the exchange rate during 1985 and 1986 was accompanied by a marked increase in interest rates in Australia, with many rates reaching record levels during this period (Figure 5.5). Much of this increase has since been reversed, with most interest rates in Australia now at levels broadly in line with those of 1984, prior to the depreciation episode.

From the viewpoint of the Australian farm sector, the real exchange rate and real interest rates are the domestic macroeconomic variables with the greatest potential to affect profitability. For example, the decline in the real exchange rate in 1985 and 1986 was so large as to cushion much of the effect of the worldwide slump in primary commodity prices at that time [Martin and Shaw (1986)]. Indeed, the key livestock industries, wool and beef, remained very profitable during this period despite the fact that measured in foreign currency the world prices of both commodities were relatively depressed. On the other hand, the very high interest rates prevailing at that time eroded part of the benefit of the low real exchange rate by adding to the cash costs of indebted farmers and by depressing farm asset values.

3. SOME KEY THEORIES

There is no question that the factors that drive exchange rates are numerous and complex. There are, however, some key theories and theoretical models that enable some order to be made out of the chaos. From a medium- and longer-term perspective, important insights can be gained from a simple theoretical paradigm developed by two Australian economists during the 1950s--Swan (1955, 1960) and Salter (1959). In the short run, the major influences on exchange

Figures 5.1 and 5.2

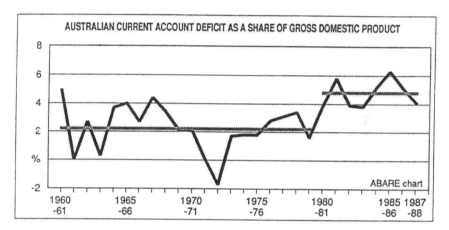

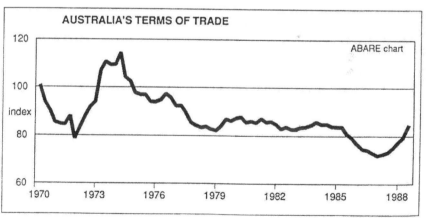

Figures 5.3 and 5.4

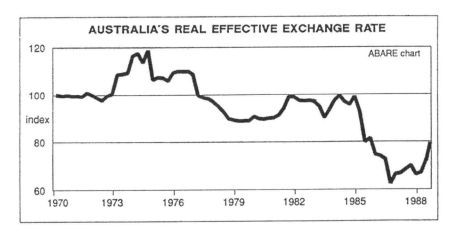

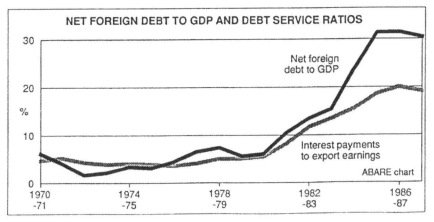

Figure 5.5

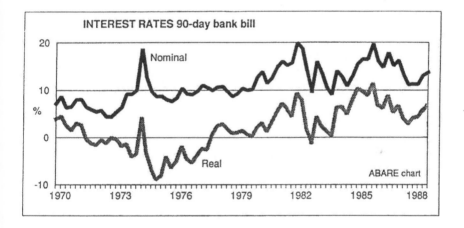

rates are likely to be monetary factors, expectations and sentiment. Theoretical and empirical work on these latter issues is now legion, but much has its origins in the seminal work by Dornbusch (1976).

In the real world, at any point in time exchange rates are driven by a combination of these longer-term and short-term influences. However, to make economic research more tractable, it is often convenient to draw a distinction between the two and focus on one or the other. This is the approach that generally has been adopted by ABARE in its exchange rate research.

3.1. The Swan/Salter Paradigm

The Swan/Salter paradigm can be conceptualized as a simple general equilibrium model that distinguishes two classes of goods and services--traded and nontraded. The key relative price in the model is the relative price of traded and nontraded goods--in other words, the real exchange rate. Since the model originated as a mode of analysis for a "small" open economy, it is assumed that the relative price of importables and exportables--the terms of trade--is determined exogenously. It is assumed that resources flow between the traded and nontraded goods sectors in response to changes in the real exchange rate. Domestic demand for both classes of goods is assumed to be a function of aggregate demand (or absorption) and the real exchange rate. A simple geometric representation of the model is set out in Appendix A, along with some standard manipulations.

A central conclusion to emerge from analysis with the model is that the role of the real exchange rate is <u>not</u> to equate the demand for and supply of traded goods, and hence to maintain equilibrium in a country's external accounts. Rather, its role is to ensure that the market for <u>nontraded</u> goods is in equilibrium.

Intuitively, the reason for this result is relatively simple. As nontraded goods cannot, by definition, be either exported or imported, then any excess demand for or supply of nontraded goods must be eliminated by substitution and switching between traded and nontraded goods. To bring this about requires a change in the relative price of traded and nontraded goods. The domestic demand for and supply of traded goods, in contrast, need not be equal at all points in time. If demand for traded goods exceeds the local production of traded goods, the difference takes the form of a balance of trade deficit and the converse.

It is also instructive to note that given equilibrium in the market for nontraded goods, an excess demand for traded goods must imply that absorption exceeds aggregate production and the converse. In other words, the driving force behind a trade imbalance is an imbalance between absorption and GNP.

The existence of a trade imbalance does not, therefore, necessarily mean that the real exchange rate is at an inappropriate level or that a change in the real exchange rate will, in isolation, provide a solution. Rather, the ultimate solution lies in the narrowing or closing of the gap between absorption and GDP. Most often this will occur more or less automatically as the private sector of the economy responds to the changes in its wealth position that have resulted from its previous borrowing and lending decisions--decisions that are likely to have been rational in the sense that they resulted in an intertemporal maximization of expected utility. However, a change in fiscal policy may be required if an excessively loose or tight fiscal policy was a major contributor to the gap between absorption and GDP, and if that fiscal policy stance did not result in a change in private sector savings in the way suggested by Barro (1974, 1979). If either or both of these adjustments occurs too slowly for the satisfaction of participants in financial markets, then they may seek to force the pace of adjustment. In the case of an external deficit, for example, investors may generate an outflow of capital, driving up domestic interest rates sufficiently to either choke off some private sector demand or to force the government to tighten fiscal policy.

As the gap between absorption and GDP narrows via one or more of these routes, then, and only then, is a sustained change in the real exchange rate likely to occur. The extent of the change in the real exchange rate will be governed by the need to maintain equilibrium in the nontraded goods sector.

The Swan/Salter model also provides important insights into the interaction, over the medium term, between a country's terms of trade and its real exchange rate. This is a crucial issue for countries (such as Australia) whose exports include a large proportion of primary commodities. Because the prices of primary commodities tend to be relatively volatile on world markets, the terms of trade of primary commodity exporters also tend to be volatile.

In the context of the Swan/Salter framework, it follows readily that the effect on the real exchange rate of a decline (for example) in the terms of trade will be governed by the change in domestic demand which accompanies that decline. If domestic demand does not decline despite the decline in real effective income brought about by the decline in the terms of trade, then the preexisting equilibrium in the nontraded goods sector may not be disturbed significantly, and hence there will be little requirement for the real exchange rate to change. On the other hand, if, as seems likely, domestic demand does decline, it is likely that the relative price of nontraded goods will fall--in other words, that a real devaluation will occur. Further, it can be shown readily that if domestic demand falls by less than the decline in real effective income following the decline in the terms of trade, the current account will deteriorate (see Appendix B). In other words, the new equilibrium real exchange rate following the decline in the terms of trade will not necessarily be associated with an unchanged balance of trade and current account outcome.

Finally, the equilibrium real exchange rate obtained from this type of analysis, focusing as it does on longer-term structural factors, can also be interpreted as a measure of the Fundamental Equilibrium Exchange Rate, or FEER, proposed by Williamson (1985). Also in keeping with the FEER approach, there is no presumption that the real exchange rate will approximate its longer-term equilibrium level at all points in time. In particular, shorter-term influences may, in some periods, cause the actual real exchange rate to differ markedly from its longer-term equilibrium level.

3.2. Theories of Short-Term Exchange Rate Behaviour

3.2.1. The Monetary Overshooting Model

Over the past decade, an extensive literature has emerged dealing with the impact of monetary policy and expectations on the nominal and real exchange rate. The primary contribution was probably that of Dornbusch (1976), with important extensions being made by, among others, Gray and Turnovsky (1979), Bhandari (1981) and Turnovsky (1981).

At the risk of some oversimplification, two key assumptions or relationships can be said to provide the driving force in these models. First, it is assumed that with capital very mobile around the world, arbitrage will ensure that interest rates, adjusted for expected exchange rate movements and risk, are equated around the world: the interest rate parity condition. Second, it is assumed that in the short term prices in financial markets such as interest rates and exchange rates are more responsive to market forces than are prices in goods markets--particularly the prices of nontraded goods. Dornbusch (1976) considered the case of a change in monetary policy under such circumstances. His results for the long run are quite conventional-- monetary policy is neutral with respect to real variables, including the real exchange rate. In other words, over the medium- to longer-term, analysts can ignore monetary influences and focus on general equilibrium or structural factors such as those identified in, for example, the Swan/Salter approach.

In the short run, however, the role of monetary influences can be much more important. A monetary shock can cause not only an immediate change in the nominal exchange rate but a change that carries it temporarily beyond its new equilibrium position--in other words, a monetary shock can cause the nominal exchange rate to "overshoot" in the short term. Further, because (by assumption) many prices in the goods market and hence the overall price level, respond to the monetary shock with a lag, the movement in the nominal exchange rate constitutes a short-run change in the real exchange rate. The relative stickiness of some prices in response to monetary shocks has been examined recently by Chalfant, et al. (1986), Devadoss and Meyers (1987) and, from an Australian perspective, by Lewis and Dwyer (1988).

The intuition behind the short-term overshooting of the nominal exchange rate is quite straightforward. Suppose, for example, that the money supply, or its rate of growth increases exogenously. With output and prices in the goods market assumed relatively fixed in the short run, the burden of maintaining equilibrium in the money market will fall on a lowering of interest rates. But at the same time, arbitrage on world financial markets will not allow the interest rate parity condition to be violated. The resolution to this apparent impasse comes in the form of an immediate downward movement in the nominal exchange rate to a point below its new long-run equilibrium position. Participants in financial markets, who are assumed to be rational (or at least quasi-rational), are thus persuaded to expect a subsequent appreciation of the nominal exchange rate and hence become willing to accept a lower interest rate. Over time, however, the price level will rise, thus eliminating the need for lower interest rates and an overshot nominal exchange rate. The combination of a rising price level and a rising nominal exchange rate will serve eventually to eliminate the initial real devaluation.

After Dornbusch's seminal work on overshooting of the exchange rate in the short run, some research effort was directed toward identifying circumstances under which an overshoot may not occur in response to a monetary shock. For example, Turnovsky (1981) argued that an overshoot may not be necessary if output is not fixed. In other words, part of the burden of maintaining equilibrium in the money market may be borne by a change in output, reducing the need for change in the interest rate and hence in the exchange rate. Gray and Turnovsky (1979) demonstrated that a pre-announcement of the monetary shock could reduce the extent of, or perhaps eliminate, the short-run overshoot of the nominal exchange rate. From an alternative perspective, Bhandari (1981) demonstrated that if the assumed instantaneous equilibrium in the money market is replaced by partial adjustment, then again a short-run overshoot in the exchange rate may be avoided.

It must be stressed, however, that the crucial issue from the viewpoint of those sectors of the economy exposed to exchange rate movements (which in most countries includes the farm sector) is not that the exchange rate may overshoot its long-run level in the short-run. It is that the exchange rate, and hence the price of traded goods, is likely to change more quickly in response to a monetary shock than will the price of nontraded goods and hence the general price level. Therefore, even if a short run overshoot is circumvented--by, for example, one or more of the "circuit-breakers" noted above--the monetary shock may well still result in a change in the real exchange rate in the short run and therefore have an effect on the farm sector.

3.2.2. Exchange Rate Expectations and Risk Premiums

The interest rate parity condition is a fundamental premise on which the overshooting model of short-run exchange rate behaviour is based. Most such models also assume that exchange rate expectations are formed rationally (or at least "quasi-rationally", or "regressively"). However, in recent papers, Frankel and Froot (1986, 1988) explored some implications of a limited departure from rationally formed exchange rate expectations and were led to the concept of a "speculative bubble." During periods when the exchange rate exhibits a significant trend, it is likely that exchange rate forecasts based on simple time series analysis would perform relatively well, and that participants in the foreign exchange market would give them a significant weighting when forming their exchange rate expectations. Such expectations may become self-fulfilling for a time, resulting in a continuation of the trend in the exchange rate beyond a level consistent with the country's longer-term economic fundamentals (such as the level that might be implied by a Swan/Salter-type analysis). However, as the time series forecasts gradually move out of line with market fundamentals, their accuracy will decline, encouraging market participants to give more weight to market fundamentals and eventually leading to a "bursting of the speculative bubble."

The theory provides little guidance as to the length of time over which a speculative bubble may persist. In the particular case of the U.S. dollar, Frankel and Froot (1986, 1988) suggest that a bubble may have been present for around 15 months, from early 1984 to March

1985. Generally, it seems reasonable to suppose that such a phenomenon would be restricted to the relatively short term.

In addition to attempting to anticipate movements in the exchange rate--perhaps on the basis of rational expectations or of the speculative bubble phenomenon--participants in financial markets may also seek a risk premium in the returns on their investments in one country relative to those in other countries. The size of that relative risk premium or discount is likely to reflect, among other factors, the degree of certainty with which the exchange rate expectation is held, as well as the more nebulous factors of optimism or pessimism about the future prospects for the economy and the likely path to be followed by economic policy.

In terms of the monetary overshooting model, the effect of the emergence of, or a change in the size of, a risk premium can be analyzed in exactly the same way as that of an exogenous change in the world interest rate. Consider the case, for example, of an increase in the risk premium. The options available to the domestic monetary authorities are essentially the same as those that would be available to them in the case of a rise in the world interest rate. On the one hand, they could tighten monetary policy sufficiently to ensure that the risk premium is incorporated rapidly into the domestic interest rate structure. As financial markets would be satisfied with the new higher domestic interest rate structure, major capital outflows would be avoided and the nominal and real exchange rates would remain relatively stable in the short run. However, to the extent that the higher domestic real interest rate reduces domestic demand for goods and services by encouraging saving and discouraging investment, some downward pressure would be placed on the real exchange rate over the medium term--an outcome that emerges readily from the Swan/Salter framework.

The alternative strategy is to leave monetary policy alone. This would mean that domestic interest rates would not rise immediately to incorporate the risk premium. In terms of the Dornbusch analysis, the outcome would be similar to the case outlined above in which the domestic money supply is assumed to increase. The nominal exchange rate would weaken immediately and "overshoot" downward, creating an expectation of a subsequent exchange rate appreciation and thus providing financial markets with the risk premium they desire. Given the assumption of sluggish prices, particularly in the nontraded goods sector, this weakening of the nominal exchange rate will also lower the real exchange rate. Over the medium term, the general price level will rise, reducing the real money stock and hence placing upward pressure on domestic interest rates. As the risk premium is reflected increasingly in the domestic interest rate structure, the need for an overshot exchange rate will be reduced. The combination of a strengthening nominal exchange rate and a rising price level will reverse much of the real devaluation that occurs in the short term. As under the first strategy outlined above, the real devaluation that remains over the medium term, if any, will be governed by the decline in demand for goods and services in response to the higher domestic real interest rate.

4. RECENT ABARE RESEARCH

4.1. Application of the Swan/Salter Theory

Perhaps the most obvious, and certainly the most popular, explanation for the dramatic decline in Australia's real exchange rate in 1985 and 1986 was the fall in the terms of trade and the associated deterioration in the current account balance and rise in the foreign debt level. For example, researchers at the Institute of Applied Economic and Social Research at the University of Melbourne [Dixon and Parmenter (1987); Fraser (1987)] argued that not only could all of the decline in the real exchange rate be explained in this way, but that even further real devaluation was warranted relative to the level reached in mid-1986. The prospect of further real devaluation also led these researchers to suggest that real interest rates would remain very high in Australia during the remainder of the 1980s as the declining real exchange rate was reflected in exchange rate expectations.

A major government advisory body, the Economic Planning Advisory Council (EPAC) has taken a slightly less extreme view. They argued that while all of the decline in the real exchange rate in 1985 and 1986 could be explained in terms of the developments in Australia's current account and terms of trade, no further decline in the real exchange rate was required relative to its mid-1986 level (EPAC 1986, 1988). With the real exchange rate expected to be relatively stable in the latter half of the 1980s, EPAC was relatively optimistic that (in contrast to the Melbourne Institute view) interest rates would decline from the levels reached in 1985 and 1986.

ABARE used as a vehicle for its analysis a large-scale general equilibrium model of the Australian economy, code named ORANI [Dixon et al. (1982)]. This model has been developed and refined over a period of well in excess of ten years by several government departments and agencies in cooperation with the University of Melbourne. It has been used extensively in industry and policy analysis during that time. Most sectors and industries in the

Australian economy are captured in some detail in ORANI. The philosophy underlying its structure is strongly neoclassical, although price rigidities and nonmarket clearance in key areas can be imposed at will.

ORANI is an excellent vehicle for quantifying the effects of the influences identified and discussed in the context of the Swan/Salter model. Indeed, to the extent that a distinction can be drawn between traded and nontraded goods in ORANI, analyses with that model can be validly thought of as an attempt to give some practical expression to the Swan/Salter theory.

To check the robustness of the results obtained from ORANI, use was also made of a much smaller general equilibrium model developed within ABARE--Crowley, O'Mara and Campbell (1983).

The first step in the ABARE research [O'Mara, Wallace and Meshios (1987); O'Mara, Crofts and Coote (1987); O'Mara (1988)] was to assess the effect on the real exchange rate of the decline in the terms of trade in 1985 and 1986. In line with the discussion of the Swan/Salter model above, a range of assumptions were made about the size of the change in absorption that would be likely to be induced by, or to otherwise accompany, the decline in the terms of trade. In brief, the results indicated that the decline in the real exchange rate would be, at most, around 10 percent.

Even before the decline in the terms of trade in 1985 and 1986, Australia's current account deficit was somewhat larger than its historical average as a share of GDP--large enough to imply that Australia's foreign debt as a share of GDP would trend upward. Australia's foreign debt to GDP ratio was already around 30 percent by 1986 and a source of some concern for policy makers and in financial markets. It therefore seemed probable that, in one form or another, adjustments would occur during the remainder of the 1980s to reduce the underlying current account deficit to a level closer to its historical average. There were, of course, several ways in which the required decline in absorption relative to GDP might come about-- increases in private saving in response to changing debt and wealth levels, a tightening of fiscal policy or, by default, enforced adjustment in the private sector in response to the imposition of a risk premium in the interest rate structure. In either case, the analysis indicated that this further adjustment would be unlikely to be accompanied by a sustained additional decline in the real exchange rate of more than around 10 percent. In other words, even if it were hypothesized that the decline in the real exchange rate in 1985 and 1986 reflected not only the concurrent decline in Australia's terms of trade but also the anticipated effects of structural adjustments likely to occur later in the

1980s, it was difficult to rationalize a fall in the real exchange rate of more than around 20 percent.

As with all general equilibrium models, the numerical results obtained must be interpreted carefully and allowance made for a significant margin of error. Nevertheless, the gap between the measured decline in the real exchange rate in 1985 and 1986--close to 40 percent--and the decline suggested by an examination of the longer-term economic fundamentals as outlined above is quite striking. It was concluded that there was plenty of scope for other shorter-term influences to have contributed to the outcome.

4.2. Application of Shorter-Term Theories

One candidate for inclusion in the list of possible short-term influences on the exchange rate during the mid-1980s is an expansionary monetary shock. A very pronounced increase occurred in the rate of growth of the major monetary aggregates in Australia in the latter part of 1984 and during 1985, around the same time as the large fall in the exchange rate (Figure 5.6). This conjunction of developments would seem to fit quite neatly into the Dornbusch overshooting framework. Unfortunately, the picture is complicated by the fact that deregulation of the Australian financial system was proceeding apace at the same time. Among other effects, deregulation improved the competitiveness of banks relative to nonbank financial institutions and led to the emergence of various new forms of financial instruments. This is likely to have caused a slowdown in the velocity of circulation of money, particularly of the more narrowly defined money supplies. Consequently, despite the fact that the monetary aggregates were observed to be growing very quickly, there was considerable uncertainty among policy makers and policy advisers as to whether this implied an excessively rapid rate of monetary growth.

With the benefit of hindsight, it now seems clear that the rate of growth of the money supply was, in fact, excessive during the latter part of 1984 and 1985. In addition to the sharp decline in the exchange rate, inflation in Australia increased sharply, from around 6 percent in 1984 to more than 9 percent in 1986, and growth in aggregate demand and GDP were very strong during much of 1985. These developments are consistent with the hypothesis that an expansionary monetary shock occurred during this period. More formal empirical support for this hypothesis has been provided by ABARE research [Hogan (1986)]. In that analysis, several conventional monetary models of short-term exchange rate behaviour were estimated using Australian data up to mid-1985. At least some

Figure 5.6

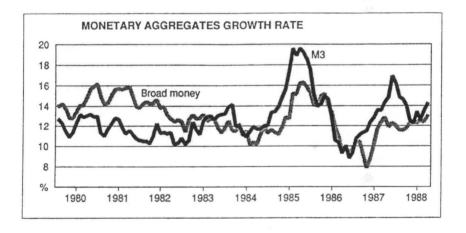

of those models seem to be capable of tracking and explaining at least part of the large decline in the exchange rate in 1985.

Other ABARE research has explored the questions of whether the emergence of a risk premium in the Australian interest rate structure or the formation of a speculative bubble in the foreign exchange market also may have contributed to the decline in the exchange rate in 1985 and 1986 [Thorpe, Hogan and Coote (1988)]. A common approach to the measurement of risk premiums is to assume that exchange rate expectations are formed rationally and hence, that given covered interest parity, the observed prediction error in the currency forward market represents the risk premium (or discount). An obvious problem with this approach is that even if currency markets are rational or speculatively efficient, prediction errors can arise as a result of new information coming to hand after the expectation is formed. In its research, ABARE adopted the alternative approach of measuring the risk premium as the difference between the forward market quote and the expected future spot rate obtained from a survey of market participants.

The results indicate that a very substantial risk premium emerged during 1985, at times reaching double digits as an annual rate. However, the risk premium seemed to become appreciably smaller during 1986, before increasing briefly in the period just after the stock market slump in October 1987. It is interesting to note that, while the risk premium appeared to become smaller in 1986, there is also evidence of less rationality in the formation of exchange rate expectations at that time. One possibility is that given the volatility of the exchange rate in 1985, survey respondents were in 1986 more inclined to base their exchange rate expectations on the forward market quote itself, thus biasing the survey results. Alternatively, the weakness in the exchange rate in 1985 may have been sufficient to create an expectation of further weakness in 1986, in the manner of a speculative bubble.

4.3. An Alternative Perspective on the Real Exchange Rate

Another key area of exchange rate research in ABARE has been the definition and measurement of the real exchange rate and of international competitiveness. This issue has generated a great deal of interest in the profession, both in Australia and elsewhere, over the past decade and has spawned a vast literature. Much of this literature has been directed at the relatively mechanistic aspects of the choice of countries and weightings in the construction of real effective exchange rate indices--particularly the construction of commodity specific real exchange rates. Many of the issues involved have been surveyed recently by Dutton and Grennes (1985, 1987) Juttner (1988) and Fagerberg (1988).

Recent ABARE research has taken as a starting point the definition of an industry's competitiveness as its ability or otherwise to attract resources from elsewhere in the economy.

From a microeconomic perspective, the key factors influencing the competitiveness of an industry thus defined would include: the output price relative to the prices of other goods in the economy; the input prices relative to those paid in other industries; the relative rate of technological change; and the relative influence of government regulations and controls. At a macroeconomic level, of course, the focus must be on broad sectors rather than on individual industries. Taking a lead from the Swan/Salter model, an obvious basis for aggregating industries is the traded and nontraded categorization. Thus, the competitiveness of traded goods in general can be defined as the ability of the traded goods sector to attract and hold resources in competition with the nontraded goods sector. The key relative price in that case is the relative price of traded and nontraded goods, or the real exchange rate.

Even when the focus is on the competitiveness of an individual industry within the traded goods sector, this approach remains valid. A sensible first step is still to consider the relative price of traded and nontraded goods in general as an indicator of the competitiveness of the overall traded goods sector, other factors unchanged. This leads naturally to the second step of considering microeconomic factors that are peculiar to the particular industry and which may make its experience different from that of other traded goods industries. An obvious such factor is a change in the price of the commodity in question on world markets relative to the prices of other traded commodities. Hence, the focus returns immediately to the familiar territory of demand for and supply of individual commodities on world markets. Many influences are likely to be at work, but included among them may be changes in the relative prices of traded and nontraded goods in other countries that are major exporters or importers of the commodity in question.

In some cases, it may be convenient to take all of these relevant real exchange rate movements and weight them together to form an index. However, that index should be interpreted as an attempt to capture the balance of exchange rate forces on the world demand and supply, and therefore on the world price, of the commodity in question--the price variable being the relevant one as far as the "competitiveness" of the local industry is concerned. Further, in view of the difficulties likely to be encountered in determining appropriate weights for each country, it may be more sensible to abandon the attempts at aggregation into an index and treat developments in each country on their merits, in much the same way as might be done for weather shocks or changes in market intervention. In any event, even if an index is to be formed, the theory outlined here indicates clearly that the variables that should be weighted together are the movements in the relative prices of traded and nontraded goods in the relevant economies rather than the more conventional real bilateral exchange rates.

Though it is common to define the real exchange rate theoretically as the relative price of traded and nontraded goods, empirical estimates generally have been obtained indirectly because of the difficulty of obtaining price data for nontraded goods. Specifically, empirical estimates generally are obtained as an index derived from the movements in the nominal exchange rates against major trading partners weighted by overall trade shares and adjusted for inflation differentials. However, the two measures are closely related, and under certain assumptions their movements will be proportional (see Appendix C). Those assumptions are that the "law of one price" holds with respect to traded goods, and that the relative prices of traded and nontraded goods do not change in any of the home country's trading partner countries (or at least that any such movements exactly cancel out). These assumptions, while plausible, are sufficiently restrictive to raise the possibility that, on occasion, the conventional empirical measure may not provide an accurate guide to the actual movements in the relative price of traded and nontraded goods. This is an important issue for ABARE because the Bureau has for many years published a regularly updated real exchange rate index computed in the conventional way, and has interpreted the movements in that index as being indicative of changes in the relative price of traded and nontraded goods [O'Mara, Carland and Campbell (1980)].

In recent ABARE research [Dwyer (1987); Dwyer and O'Mara (1988)] an attempt has been made to obtain a direct measure of movements in the relative price of traded and nontraded goods in Australia. Implicit price deflators are readily available for exports and imports, and it was assumed that the prices of exportables consumed domestically and importables produced domestically would move in sympathy. Making a range of plausible assumptions about the share of traded goods in the CPI, the implied movements in the price of nontraded goods were obtained by deduction. Finally, to remove the direct price effects of changes in Australia's terms of trade, it was assumed that movements in export prices were identical to the movements in import prices. The ABARE's conventional real exchange rate measure and this alternative measure are shown in Figure 5.7. The relative price index is represented as a band to reflect the use of a range of assumptions about the share of traded goods in the CPI. A fall in that locus implies a rise in the relative price of traded goods, and the converse.

While correlation between the two series is evident, the relationship is obviously far from precise. Of particular interest are the movements in 1985 and 1986. As noted earlier, the conventional measure of Australia's real exchange rate declined by close to 40 percent during that period. However, the rise in the relative price of traded goods, though substantial, was significantly less at around 20 percent. Several factors are likely to have been at work in producing this divergence. The real exchange rates of several of Australia's trading partners changed significantly during this period--most notably those of the United States and Japan. However, as the movements were in opposite directions, their effects on the conventional real exchange rate would have cancelled out to some extent. Following such a major change in the exchange rate, the "law of one price" is also likely to have been violated in the short term as foreign suppliers of imports absorbed some of the effects of the devaluation to maintain market share. Contractual arrangements also may have slowed the flow-on of the devaluation into the domestic price of traded goods.

It is noteworthy, however, that over the period since the latter part of 1986, during which time the Australian dollar has trended upward, the relative movements in the two series have been much more comparable. In other words, the effects of contractual arrangements, departures from the law of one price and so on seem to have produced much less tardiness in the price of traded goods in the presence of an appreciation than during the earlier devaluation episode. Consequently, while the real effective exchange rate remains about 20 percent below its level of 1984, the relative price series has returned to around its 1984 level--although it remains to be seen whether subsequent data will result in some modification of this conclusion. Nevertheless, should the relative price series remain around its 1984 level, it may indicate that some real devaluation is in prospect, given that some sustained improvement in competitiveness is required as part of the process of reducing Australia's current account deficit and foreign debt burden.

More generally, it may be sensible to interpret the conventional measure of the real exchange rate as providing some indication of the change that might occur <u>ultimately</u> in the relative prices of traded and

Figure 5.7

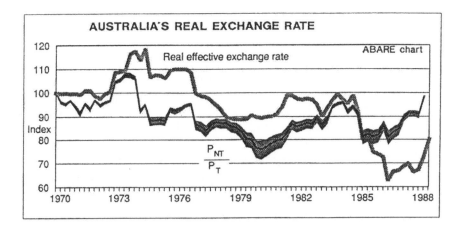

nontraded goods after sufficient time has elapsed for the various lags and rigidities to have been worked out.

4.4. Summing Up

It may be useful to now draw together the various strands of the recent ABARE research in the area of exchange rates and interest rates and to assess the extent to which it has enabled ABARE to place a quite specific interpretation on recent exchange rate and interest rate developments in Australia. First, analysis of the medium- and longer-term influences on the exchange rate along the lines of the Swan/Salter model led ABARE to conclude that the dramatic decline in the real exchange rate in 1985 and 1986 was excessive and was probably driven in part by unsustainable shorter-term influences. Other ABARE research has helped fill out some of the details of these shorter-term influences. In particular, there is evidence of an expansionary monetary shock in the spirit of the Dornbusch overshooting analysis, evidence of the emergence of a risk premium in the Australian interest rate structure, and some evidence of a "speculative bubble" in the foreign exchange market. Finally, there is evidence that the price of traded goods relative to those of nontraded goods increased less than might have been anticipated following the devaluation, at least partly due to various lags, although these lags seem to have been less significant during the period of exchange rate appreciation since the latter part of 1986.

Some uncertainty remains as to why the results of the ABARE research differ so markedly from those of Dixon and Parmenter (1987) and Fraser (1987) and, to a lesser extent, EPAC (1986, 1988). It will be recalled that these authors argued that all the developments on the exchange rate and interest rate fronts in 1985 and 1986 could be explained without appeal to short term influences. Dixon and Parmenter also argued that interest rates would remain very high over the remainder of the 1980s and that the real exchange rate would fall even further from its level in 1986. Use of different models provide part of the explanation, although the Dixon and Parmenter analysis also used a version of ORANI.

A more important explanation probably lies in the assumptions used, particularly with respect to fiscal policy. In the ABARE research, it was assumed that fiscal policy would be tightened over the remainder of the 1980s sufficiently to reduce the current account deficit to a level consistent with its longer-term average level--a level that would also allow Australia's foreign debt to GDP ratio to be stabilized. Dixon and Parmenter assumed a similar outcome for the current account balance but also assumed that no contribution would be made by a tightening of fiscal policy. Rather, all of the restraint on absorption was assumed to occur in the private sector in response to very high interest rates that, in turn, required a rapidly devaluing currency to maintain the interest rate parity condition. However, even here a major puzzle remains because the cumulative effects of such projected real devaluations would carry the real exchange rate to a much lower level by around 1990 than was found necessary by any other researchers.

Armed with its own research results, ABARE has, since the latter part of 1986, based all of its assessments of the medium-term outlook for commodity prices in Australia on an assumption that the real exchange rate would trend upward significantly. It was also assumed that interest rates in Australia would decline significantly from the very high levels ruling in 1985 and 1986 as the risk premium in the interest rate structure declined and as the appreciating trend in the real exchange rate was incorporated into expectations. To date at least, both of these assumptions have proven to be quite close to the mark. Almost half the real devaluation in 1985 and 1986 has been reversed, and nominal and real interest rates have declined significantly since the latter part of 1985. Certainly there is no evidence of the Dixon and Parmenter projections coming to pass. ABARE's latest assessment, based on updates of the earlier research and taking into account the recent recovery in Australia's terms of trade, is that scope remains over the medium term for further upward movement of the real exchange rate, as conventionally measured, and downward movement in real interest rates. However, the conclusion with respect to the exchange rate may need to be modified if the unexpectedly strong recovery in the relative price of nontraded goods since 1986 is maintained.

Finally, it is clear that there are some similarities and some differences between the Australian experience since the mid-1980s and the U.S. experience, particularly in the first half of the 1980s. In both cases, there is evidence that monetary policy contributed to the large change in exchange rates, albeit in opposite directions, in the two countries. The U.S. experience is well documented in, for example, Chalfant et al. (1986). In both cases, too, a "speculative bubble" may have extended the movement in the exchange rate; the U.S. evidence is outlined by Frankel and Froot (1986). On the other hand, it is clear that the other major contributor to the U.S. experience was another policy variable, namely expansionary fiscal policy, while in the Australian case it was a largely exogenous deterioration in the terms of trade. The role of fiscal policy in the real devaluation of the

Australian dollar in 1985 and 1986 is problematical. Fiscal policy was, in fact, relatively expansionary at that time, pointing to upward pressure on the real exchange rate, in line with the U.S. experience, rather than downward pressure. However, the large budget deficits at that time may have contributed to the emergence of a risk premium in the Australian interest rate structure, which may have been sufficient to reverse the sign on the relationship between fiscal policy and the exchange rate.

5. SOME INTERNATIONAL ISSUES

Large current account imbalances and major swings in real exchange rates (decrease indicates devaluation) have been key features of the world economy in recent years, and are likely to remain the crucial issues for policy makers in the foreseeable future (Figures 5.8 and 5.9). In that sense, Australia's experience has been something of a microcosm of developments in the world economy.

General equilibrium theory in the spirit of the Swan/Salter analysis indicates clearly that if the current account imbalances in the major economies are to be reduced to levels more acceptable to financial markets, changes may be required to both absorption and real exchange rates. Changes in real exchange rates in isolation may not be successful or even sustainable over the medium term without an appropriate change in absorption. This is a simple point, but it did not appear to be recognized fully at the time of the Plaza Agreement in September 1985, where the focus was mainly on the perceived need to achieve a major devaluation of the U.S. dollar and an appreciation of the Yen and the Deutschmark following the marked increase in the real value of the U.S. dollar in the first half of the 1980s. More recently, the need for a two-pronged approach has become much more widely recognized, possibly reflecting in part the slowness of the current account imbalances to decline in the face of the very large exchange rate movements that have occurred since 1985.

There is also now widespread agreement that a decisive reduction in the U.S. budget deficit is central to the changes in absorption that are required. Much has been written on this issue recently--for example, Branson (1986), Dornbusch (1986, 1987). However, a decline in savings relative to investment in the U.S. private sector during the 1980s has also made an important contribution to the deterioration in the current account balance. This stands out clearly in some estimates compiled recently by ABARE [Wallace, Bramma and O'Mara (1988)] --see Table 5.1. It remains to be seen whether the wealth effects of this imbalance between saving and investment will produce a

Figures 5.8 and 5.9

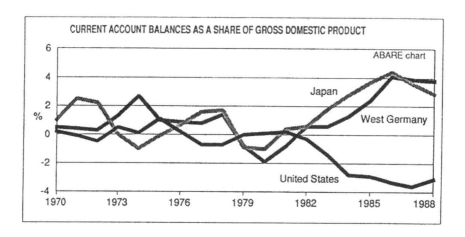

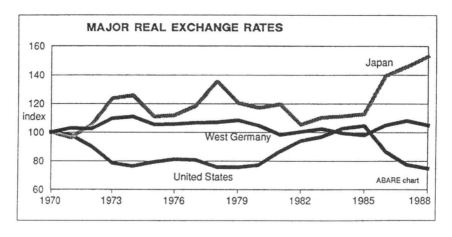

sufficiently powerful self-correcting mechanism in the private sector over the next fewyears, or whether financial markets will seek to force the pace of adjustment by imposing a risk premium on the US interest rate structure. Even if financial markets are relatively unconcerned about the savings/investment imbalance in the private sector per se, they may still force adjustment pressure on to the private sector in the form of a risk premium if a further tightening of US fiscal policy is not forthcoming.

There is also some uncertainty about the most appropriate changes, if any, to fiscal policy in Japan and the Federal Republic of Germany. Though fiscal policy has been tightened significantly in both countries during the 1980s, this change commenced from a position of large budget deficits, and even now both countries still have modest budget deficits (see Tables 5.2 and 5.3). It would seem, therefore, that the case for the adoption of much more expansionary fiscal policies in these countries for an extended period is much weaker than is the case for a tightening of fiscal policy in the United States. The large current account surpluses in Japan and West Germany largely reflect an excess of savings over investment in the private sector, and any sustained reduction in the current account surpluses will therefore probably require adjustments in the private sector. A reduction in world real interest rates in response to a decline in the demand for capital inflow into the United States in future years is one possible source of such a change.

While it is clear that the major economies will require further adjustments to absorption over the next few years if the current account imbalances are to be reduced to levels more acceptable to financial markets, it is much less certain that further adjustments will be required to their real exchange rates. Perhaps the key lesson from exchange rate theory is that relatively short-term influences on exchange rates can be quite pervasive and can carry the exchange rate temporarily to a level well removed from that which is sustainable over the longer term. Large adjustments have occurred since 1985 in the value of the U.S. dollar, the Yen and the Deutschmark, probably induced in part by monetary and speculative influences. Even if further changes are required, they will not necessarily be in the directions that might be suggested by intuition--that is, downward in the case of the U.S. dollar and upward in the case of the Yen and the Deutschmark. For example, it is possible in principle that the recent exchange rate adjustments may have contained an element of overshooting, much as seems to have occurred in Australia in 1985 and 1986. Certainly, the mere observation that the current account imbalances remain substantial does not provide conclusive proof of the

TABLE 5.1. United States: Gross Saving and Investment, by
            Sector

Proportion of gross national product

Year	Public investment[a]	Private investment	Public saving[b]	Private saving	Current account balance
	%	%	%	%	%
1973	0.0	18.1	0.6	18.0	0.5
1974	0.6	16.3	-0.3	17.3	0.1
1975	0.1	13.7	-4.1	19.0	1.1
1976	0.0	15.6	-2.2	18.0	0.2
1977	0.2	17.3	-1.0	17.8	-0.7
1978	0.4	18.5	0.0	18.2	-0.7
1979	0.2	18.1	0.5	17.8	0.0
1980	0.1	16.0	-1.3	17.5	0.1
1981	-0.1	16.9	-1.0	18.0	0.2
1982	0.3	14.1	-3.5	17.6	-0.3
1983	0.3	14.7	-3.8	17.4	-1.4
1984	0.3	17.6	-2.8	17.9	-2.8
1985	0.2	16.0	-3.3	16.6	-2.9
1986	0.2	15.7	-3.5	16.1	-3.3
1987[c]	0.2	15.7	-2.4	14.7	-3.6
1988[d]	0.2	16.3	-2.3	15.7	-3.1

[a]Imputed as a residual (equal to public and private sector savings less current account balance and private investment).

[b]General government financial balances.

[c]Preliminary estimate. [d]Forecast.

SOURCE: Wallace, Bramma and O'Mara (1988).

TABLE 5.2. Japan: Gross Saving and Investment, by Sector

Proportion of gross national product

Year	Public investment	Private investment	Public saving[a]	Private saving[b]	Current account balance
	%	%	%	%	%
1973	9.6	26.8	0.6	35.8	0.0
1974	9.0	25.8	0.4	33.4	-1.0
1975	9.1	23.4	-2.7	35.1	-0.1
1976	8.7	22.5	-3.7	35.6	0.7
1977	9.1	21.1	-3.8	35.6	1.6
1978	9.8	20.6	-5.5	37.6	1.7
1979	9.9	21.7	-4.7	35.4	-0.9
1980	9.5	22.1	-4.4	35.0	-1.0
1981	9.5	21.3	-3.8	35.0	0.4
1982	9.0	20.7	-3.6	33.9	0.6
1983	8.4	19.1	-3.7	33.0	1.8
1984	7.7	20.1	-2.1	32.7	2.8
1985	6.8	20.9	-0.8	32.2	3.7
1986	6.7	20.9	-1.1	33.1	4.4
1987[c]	6.9	21.9	-0.2	32.6	3.6
1988[d]	7.2	23.4	-0.3	33.8	2.9

[a]General government financial balances.

[b]Imputed as a residual (equal to the sum of private and public investment less public saving plus the current account balance).

[c]Preliminary estimate.

[d]Forecast.

SOURCE: Wallace, Bramma and O'Mara (1988).

TABLE 5.3. FR Germany: Gross Saving and Investment, by Sector

Proportion of gross national product

Year	Public investment	Private investment	Public saving[a]	Private saving[b]	Current account balance
	%	%	%	%	%
1973	3.8	20.0	1.2	23.9	1.3
1974	4.1	17.5	-1.3	25.6	2.7
1975	3.9	16.5	-5.7	27.1	1.0
1976	3.5	16.5	-3.4	24.3	0.9
1977	3.3	17.0	-2.4	23.5	0.8
1978	3.3	17.3	-2.4	24.4	1.4
1979	3.4	18.3	-2.5	23.4	-0.8
1980	3.6	19.0	-2.9	23.6	-1.9
1981	3.2	18.5	-3.7	24.6	-0.8
1982	2.8	17.6	-3.3	24.3	0.6
1983	2.5	18.0	-2.5	23.6	0.6
1984	2.4	17.7	-1.9	23.3	1.3
1985	2.3	17.2	-1.1	23.0	2.4
1986	2.4	16.9	-1.2	24.6	4.1
1987[c]	2.4	16.8	-1.7	24.8	3.9
1988[d]	2.4	17.2	-2.6	26.0	3.8

[a]General government financial balances.

[b]Imputed as a residual (equal to the sum of private and public investment less public saving plus the current account balance).

[c]Preliminary estimate.

[d]Forecast.

SOURCE: Wallace, Bramma and O'Mara (1988).

need for further exchange rate adjustment in the conventional direction.

In the light of that conclusion, it is perhaps not surprising that opinion seems to remain divided about the prospects for the key international currencies over the medium term. For example, Dornbusch (1987) argued that the U.S. dollar might need to decline much further than it has done to date, to around 100 Yen. However, other eminent commentators in this field, such as McKinnon and Williamson, recently have been reported to hold the contrary view that the U.S. dollar is at present substantially undervalued [Economist (1988)].

There are grounds for supposing that the real levels of the U.S. dollar and the Deutschmark in the September quarter 1988 may have been broadly consistent with, or slightly above, their sustainable longer-term levels, while the yen may have been somewhat overvalued in real terms. The real value of the U.S. dollar in the September quarter 1988 was around 5 percent below its level of the early 1980s, the most recent period in which the U.S. economy was close to internal and external balance. Since that time, however, the United States has moved from being a net creditor nation to a net debtor nation, and by the early 1990s the turnaround in net transfers abroad could be equivalent to around 0.7 percent of GDP (see Tables 5.4 and 5.5). Assuming that the overall current account deficit has been reduced to more acceptable levels by the early 1990s, then the balance of trade must be sufficiently stronger than in the early 1980s to allow these debt servicing costs to be met. This means, of course, that absorption must be lower relative to GNP than it was in 1980, and with other factors unchanged the real exchange rate also must be lower than in 1980. On the basis of plausible assumptions about the relationship between absorption and the real exchange rate, as set out in Table 5.4, a sustained decline in the real exchange rate of between approximately 1 to 5 percent relative to its level in the early 1980s may be required, other factors unchanged.

Adopting a similar line of analysis with respect to the Yen and the Deutschmark, it is likely that the sustainable real value of both currencies is somewhat higher than in the early 1980s, the most recent period in which Japan and West Germany were in approximate internal and external balance. In the interim, both countries have experienced a large buildup in their net foreign assets, and this process is likely to continue for several more years yet (see Tables 5.4 and 5.5). The net income received from these investments, if reflected in a higher level of absorption relative to GNP than was the case in the

TABLE 5.4. Expected Changes in the Net External
Asset Positions of the United States, Japan and
Germany[a]

	Nominal GNP		Current account balance		Net external asset position[b]	
	Change[c]	Level	Level	Share of GNP	Level	Share of GNP
	%	US$[b]	US$[b]	%	US$[b]	%
U.S.						
1980	8.9	2 731.9	1.9	0.1	95	3.5
1987	6.7	4 526.7	-154.0	-3.4	-435	-9.6
1988[d]	6.0	4 798.3	-141.1	-2.9	-577	-12.0
1989[d]	6.5	5 110.2	-133.8	-2.6	-710	-13.9
1990	7.0	5 467.9	-109.4	-2.0	-819	-15.0
1991	7.5	5 878.0	-76.4	-1.3	-896	-15.2
1992	7.5	6 318.8	-44.2	-0.7	-940	-14.9
1993	7.5	6 792.8	0	0	-940	-13.8
		Yen[b]	Yen[b]		Yen[b]	
Japan						
1981	7.0	256 817	1 052	0.4	2 177	0.8
1987	4.1	344 888	12 416	3.6	32 852	9.5
1988[d]	6.0	365 581	9 871	2.7	42 723	11.7
1989[d]	6.0	387 516	9 688	2.5	52 411	13.5
1990	6.0	410 767	7 805	1.9	60 216	14.7
1991	6.0	435 413	5 660	1.3	65 876	15.1
1992	6.0	461 538	2 769	0.6	68 645	14.9
1993	6.0	489 230	0	0	68 645	14.0

Table 5.4 (continued)

	Nominal GNP		Current account balance		Net external asset position[b]	
	Change[c]	Level	Level	Share of GNP	Level	Share of GNP
	%	US$[b]	US$[b]	%	US$[b]	%
		DM[b]	DM[b]		DM[b]	
Germany						
1981	4.0	1 545.1	-7.7	-0.5	45	2.9
1987	3.8	2 023.2	80.8	4.0	239	11.8
1988[d]	3.0	2 083.9	68.8	3.3	308	14.8
1989[d]	4.0	2 167.3	67.2	3.1	375	17.3
1990	5.0	2 275.6	52.3	2.3	527	18.8
1991	5.0	2 389.4	38.2	1.6	466	19.5
1992	5.0	2 508.9	20.1	0.8	486	19.4
1993	5.0	2 634.3	0	0	486	18.4

[a]1990-93 figures are ABARE estimates.

[b]Defined as the sum of overseas lending and direct investment by the home country less the sum of lending and direct investment by foreigners in the home country.

[c]From previous year; ABARE estimates for 1988-93.

[d]Levels of the current account balance and net external assets are IMF estimates.

TABLE 5.5.   Expected Changes in Real Exchange Rate
             for the United States, Japan and
             Germany Between September Quarters 1988
             and 1993

	United States	Japan	West Germany
Percentage point change in net external assets to GNP ratio (1980-93)[a]	-17.3	13.2	5.5
Percentage point change in net external income to GNP ratio (1980-93)[a],[b]	-0.7	0.5	0.6
Percentage change in terms of trade (1980-93)[c]	-3.1	13.1	4.4

Table 5.5 (continued)

Required percentage change in real
exchange rate between Sept quarter
1988 and 1993, assuming that external
balance is re-established in the
presence of the changes outlined
above, taking into account the real
exchange rate changes which have
occurred since 1980[d] and assuming
the following elasticities[e]

Rise in real rate associated with 1% rise in terms of trade	Fall (rise) in the real rate associated with 1% fall in trade deficit (surplus) to GNP ratio[f]	US	Japan	W. Germany
%	%	%	%	%
0.2	2	-0.0	-17.4	-5.8
	5	-2.1	-16.2	-4.0
	8	-4.2	-14.9	-2.3
0.5	2	-1.0	-14.3	-4.5
	5	-3.1	-13.0	-2.8
	8	-5.2	-11.8	-1.1
0.8	2	-1.9	-11.2	-3.3
	5	-4.1	-9.9	-1.6
	8	-6.2	-8.7	0.1

[a]Based on figures in table 4. 1981-93 for Japan and Germany.
[b]Due to the change in the net external asset position. Assumes
a 4 percent real return on assets.
[c]Assumes a 10 percent fall in 1988 and no change thereafter.
1981-93 for Japan and Germany.
[d]Based on real exchange rates published by Morgan Guaranty.
[e]In both cases, a relatively wide range of elasticities has
been used, the midpoint of the range being broadly consistent
with the Australian experience as outlined in, for example,
O'Mara, Wallace and Meshios (1987).
[f]The fall in the trade balance to GNP ratio is assumed to be
equal, but opposite in sign, to the change in the net external
income to GNP ratio over the period from 1980 to 1993.

early 1980s, may require a real exchange rate approximately 1 to 5 percent higher than in that earlier period.

It is also important to allow for changes in the terms of trade experienced by these countries over this period. All three countries enjoyed a rise in their terms of trade between the early 1980s and 1987, with Japan experiencing the largest rise. The recent recovery in commodity prices may see some reversal of that trend. If, in all three countries, the terms of trade are assumed to fall by 10 percent in 1988 and then to remain stable at that new level over the medium term to 1993, the U.S. terms of trade would be about 3 percent below the level of the early 1980s (see Table 5.4). That might provide some justification for a real exchange rate perhaps 0.5 to 2.0 percent below that of the early 1980s. The same assumptions imply that Japan's terms of trade would remain around 13 percent above and W. Germany's around 4 percent above their levels in the early 1980s--in both cases pointing to some upward pressure on the real exchange rate.

Adding together the effects of changes in net foreign assets and terms of trade since the early 1980s (see table 5.5), the equilibrium real level of the U.S. dollar may be, at most, around 6 percent below that of the September quarter 1988. In the case of the yen, the equilibrium real level could be at least 8 percent and perhaps closer to 20 percent below the September quarter 1988 level. The situation for the Deutschmark is similar to that for the U.S. dollar.

Such estimates are, of course, only approximations. Apart from the uncertainties about the size of the elasticities, it is possible that structural changes within each economy have also had some influence on the equilibrium real exchange rates. For example, the traded and nontraded goods sectors may have had unequal rates of growth in capital stocks and in technological advancement. Thus, in terms of the geometric representation of the Swan/Salter model in Appendix A, the shape of the production possibility frontier may have changed over time. Further, in light of the Australian experience, it would be important to assess the extent to which movements in the relative price of traded and nontraded goods have mirrored the movements in the conventionally measured real exchange rate index during the 1980s.

6. SUMMARY AND CONCLUSIONS

The Australian experience of the mid-1980s is in one respect similar to the U.S. experience during the first half of the 1980s: It seems to have provided another practical illustration of the phenomenon of overshooting in financial markets. In the mid-1980s, a sharp decline in Australia's terms of trade and a deterioration in the current account deficit and foreign debt level provided a sound theoretical rationale for a significant decline in the exchange rate. However, ABARE research indicates that such factors alone are unlikely to explain the full extent of the decline in the exchange rate or the very high real interest rates that emerged at the same time. Shorter-term factors, including an element of overshooting in response to an expansionary monetary shock, a risk premium in the Australian interest rate structure, and possibly a "speculative bubble" in the foreign exchange market, also are likely to have been important. The marked recovery in the exchange rate and fall in interest rates in 1987 and 1988 probably reflect, at least in part, a gradual dissipation of these shorter-term influences, allowing the earlier overshooting to be reversed. This is not to deny that the recovery that has occurred in Australia's terms of trade since early 1987 has also played a role.

Other researchers, however, have reached different conclusions, arguing that all of the fall in the exchange rate and the rise in interest rates can be explained by the deterioration in the terms of trade and the need to achieve a satisfactory current account balance. While these differences among conclusions are partly attributable to differences in the models used, the assumptions made with respect to fiscal policy also seem crucial. In the ABARE research, it was assumed that fiscal policy would be tightened sufficiently to complement whatever adjustments occurred in the private sector so as to ensure that the outcome for the current account balance would be acceptable to financial markets. The alternative assumption--seemingly adopted by other researchers--is that fiscal policy would not be tightened and hence that financial markets would force additional adjustment on the part of the private sector via the maintenance of very high interest rates and, in the short term at least, a very low exchange rate.

Other ABARE research has focused on the relationship between the traditional empirical measure of the real exchange rate and its theoretical analogue, the relative price of traded and nontraded goods. It was found that the relative price of traded goods did not rise, at least not immediately, by as much as might have been expected from the extent of the decline in the traditional real exchange rate in 1985 and 1986; all of that increase subsequently was reversed. This must be borne in mind when assessing the extent of the overshoot in the exchange rate.

In the international arena, resolution of the current account imbalances of the major industrial economies requires, in principle, adjustments to both absorption and real exchange rates. The adjustments that have occurred in absorption to date have been insufficient for that purpose, particularly with respect to fiscal policy in the United States. In contrast, there is some evidence that the major realignment of real exchange rates that has occurred since 1985 may have brought those exchange rates--particularly those of the U.S. dollar and the Deutschmark--quite close to their sustainable longer-term levels. In other words, as in Australia in 1985 and 1986, these exchange rate adjustments may have occurred much more quickly than the adjustments to absorption. However, unlike the Australian case, it is not clear that the exchange rate adjustments contained a large element of short-run overshooting. With the possible exception of the Yen, the exchange rates may have been carried close to--rather than markedly beyond--their longer-term equilibrium levels. Even so, this dichotomous adjustment of exchange rates and absorption is unlikely to be viable except in the relatively short term. In other words, unless the adjustments to absorption catch up relatively soon, it is likely that market forces will prevent the present levels of the real exchange rates from being sustained until such time as absorption has adjusted.

Finally, there is a range of issues on which further research by ABARE and others may prove useful. One example is the factors that determined the size and behavior of the apparent risk premium in the Australian interest rate structure in the mid-1980s, including the role of fiscal and monetary policy. A second is the relationship between the alternative measures of the real exchange rate, particularly the size and nature of the lags involved. Extending that analysis to the major international exchange rates may also help to clarify whether the adjustments to these exchange rates since 1985 have carried them to a level broadly consistent with a longer-term equilibrium.

APPENDIX A: THE SWAN/SALTER MODEL

The model developed by Swan (1955, 1960) and Salter (1959) is illustrated in Figure A1. The domestic economy is assumed to be small and open, and comprised of a traded and a nontraded goods sector. The prices of traded goods are determined on world markets, and hence terms of trade is exogenous. The price of nontraded goods is determined by demand and supply factors in the domestic economy.

Quantities of traded goods are measured on the horizontal axis, and of nontraded goods on the vertical axis. The production possibility frontier (PP) represents all combinations of traded and nontraded goods that can be produced efficiently. Indifference curves such as II represent combinations of traded and nontraded goods, among which consumers have equal preference.

Suppose that initially the relative price of traded and nontraded goods is represented by the slope of the lines EF and CD. The point of tangency between EF and PP defines the initial level of output of nontraded goods, N_s, and traded goods, T_s. Also suppose, however, that the initial level of aggregate expenditure is represented by the line CD. The point of tangency between CD and II defines the initial level of demand for nontraded goods, N_D, and traded goods, T_D.

It is clear that, as drawn, the initial situation is one of excess demand for both traded and nontraded goods. The excess demand for traded goods will be expressed in a balance of trade deficit--a situation that can be sustained for some time. The excess demand for nontraded goods, in contrast, clearly is not sustainable and will engender a rise in the price of nontraded goods relative to traded goods--that is, a rise in the real exchange rate.

The rise in the relative price of nontraded goods changes the slope of the expenditure and income lines, as in Figure A2. The effect is an increase in the production of nontraded goods and a fall in demand for them, so that equilibrium is reached in the nontraded goods market. Conversely, production of traded goods falls and demand increases, so that the balance of trade deficit <u>deteriorates</u>. In other words, the role of the change in the real exchange rate has been to re-establish equilibrium in the nontraded goods sector rather than in the traded goods sector. The higher real exchange rate has reduced the ability of the traded goods sector to hold resources in competition with the nontraded goods sector--in other words, the competitiveness of the traded goods sector has been reduced.

In Figure A2, the driving force behind the balance of trade deficit is clearly the excess of demand over income. While that gap remains, it will not be possible to reduce the balance of trade deficit simply by raising the relative price of traded goods. For example, in Figure A3, the relative price of traded goods has been raised to a level at which, ceteris paribus, production of and demand for traded goods are equated but only at the expense of increased excess demand for nontraded goods--an unsustainable condition.

Changes in the terms of trade are a little less convenient to analyze geometrically. One approach is to add a third dimension to the diagram to allow traded goods to be separated into importables and exportables. It is more common, however, to retain the two dimensional representation and to recognize that a change in the terms of trade can be expressed by a biased movement in a "shadow" production possibility frontier.[2] Consider the case, for example, of a rise in the terms of trade. For any given level of output of nontraded goods, the economy now has the potential to consume a larger bundle of traded goods than previously without running a balance of trade deficit. This is similar, in effect, to an increased ability to produce traded goods. In Figure A4, the economy is initially in internal and trade balance. PP is the actual production possibility frontier; PP_1, the shadow frontier, is PP with the improvement in the terms of trade factored into the output of traded goods. Assuming that the <u>average</u> price of traded goods is unchanged following the rise in the terms of trade, and hence that the real exchange rate does not change, output is determined at point A, but effective real income is given by point B on PP_1.

Provided that at least part of the rise in real effective income is reflected in increased absorption, then the expenditure line will also move to the right, from EF to (say) GH. If the expenditure elasticity of demand for nontraded goods is positive, it is clear that this rightward movement of the expenditure line will generate excess demand for nontraded goods, placing upward pressure on the relative price of nontraded goods. The extent of that upward movement will be governed by the extent of the rightward movement of the expenditure line--in other words, by the rise in absorption that is induced by the rise in the terms of trade.

Figure A1

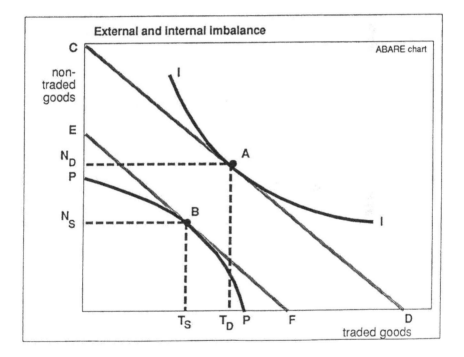

Figure A2

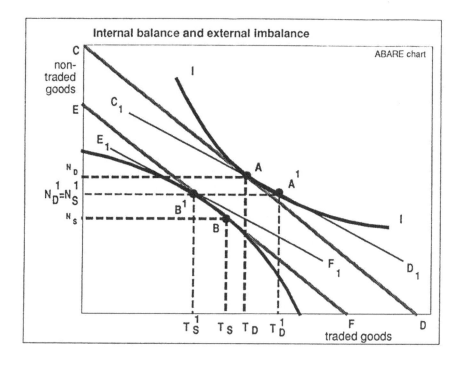

Figure A3

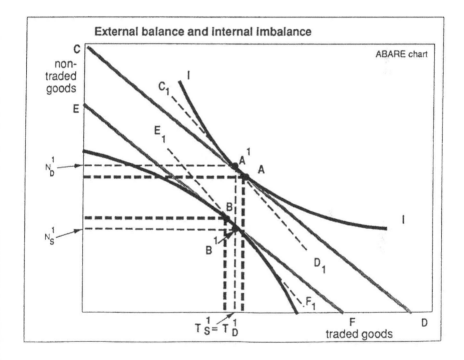

Figure A4

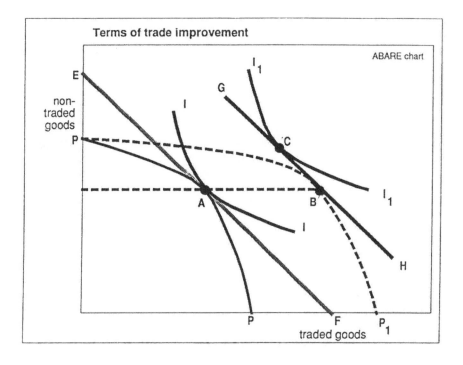

APPENDIX B: TERMS OF TRADE, ABSORPTION AND THE BALANCE OF TRADE

The relationship between the terms of trade, absorption and the balance of trade can be considered in the context of the following very simple model.

Notation

```
Y = real GDP
C = aggregate volume of consumption expenditure
I = aggregate volume of investment expenditure
G = aggregate volume of government expenditure
X = aggregate volume of exports
M = aggregate volume of imports
```
p_i = implicit deflator for i, where i = y, c, i, g, x and m to indicate GDP, consumption, investment, government spending, exports and imports respectively.

Model

The national accounting identity can be expressed in nominal terms thus:

$$p_y Y = p_c C + p_i I + p_g G + p_x X - p_m M \qquad (B1)$$

For simplicity, it is assumed that the volume of consumption expenditure is a simple function of the effective spending power of GDP:

$$C = a + b \frac{p_y Y}{p_c}, \quad b > 0 \qquad (B2)$$

Similarly, there is a positive relationship between the volume of investment expenditure and real GDP expressed in terms of investment goods. To the extent that a rise (for example) in the terms of trade raises p_y relative to p_i, it would stimulate investment, and the converse. Thus:

$$I = c + d \ \frac{p_y Y}{p_i}, \quad d > 0 \tag{B3}$$

The volume of imports is a function of the real level of consumption and investment expenditure when expressed in terms of importables:

$$M = e + f \ (\frac{p_c C + p_i I}{p_M}), \quad f > 0 \tag{B4}$$

For simplicity, government expenditure and the volume of exports are assumed to be fixed:

$$G = \bar{G} \tag{B5}$$

$$X = \bar{X} \tag{B6}$$

Analysis

Solving for M and taking the first partial derivative with respect to p_x,

$$\frac{\delta M}{\delta p_x} = \frac{f(b + d) \ \bar{X}}{p_m [1 + (b + d)(f - 1)]} \tag{B7}$$

Suppose that $b + d = 1$. In other words, suppose that a change in $p_y Y$ induces an equivalent combined change in consumption and investment expenditure. Then:

$$\frac{\delta M}{\delta p_x} = \frac{\bar{X}}{\delta p_m} \tag{B8}$$

Defining the balance of trade, BT, to be:

$$BT = p_x \bar{X} - p_m M \tag{B9}$$

then

$$\frac{\delta BT}{\delta p_x} = 0 \tag{B10}$$

Hence, in this case, the balance of trade is invariant to a change in export prices.
Conversely, if b + d = 0, then:

$$\frac{\delta BT}{\delta p_x} = \bar{X} \tag{B11}$$

and hence:

$$\delta BT = dp_x \bar{X} \tag{B12}$$

In other words, in this case a change in the export prices is reflected fully in the balance of trade.

APPENDIX C: RELATIONSHIP BETWEEN
REAL EXCHANGE RATE MEASURES

Suppose that the home country has n trading partners. In each of those n countries, the change in the CPI can be decomposed into the change in nontraded goods prices and the change in traded goods prices.

$$\Delta CPI^i = \alpha^i \Delta P^i_{NT} + (1 - \alpha^i) \Delta P^i_T, \quad i = 1...n \qquad (C1)$$

where ΔCPI^i is the percentage change in the CPI in country i, $\Delta P^i_N T$ and ΔP_T are the percentage changes in nontraded and traded goods prices in country i and α^i is the share of nontraded goods in the CPI in country i.

Assuming that the law of one price holds with respect to traded goods, then

$$\Delta P^H_T = \sum_{i=1}^{n} \lambda^i \Delta(P^i_T e^i) \qquad (C2)$$

That is, the percentage change in the price of traded goods in the home economy, ΔP^H_T, is equal to the weighted sum of the percentage changes in traded goods prices in all of the home country's trading partners adjusted for changes in the exchange rate, e, between the home country's currency and that of each of its trading partners (defined as the number of units of the home country's currency per unit of foreign currency). The weight assigned to country i, λ^i, is the share of the home country's total trade which is conducted with country i.

Assuming that none of the home country's trading partners experience a change in the relative price of traded and nontraded goods, then:

$$\Delta P^i_T = \Delta P^i_{NT} = \Delta CPI^i, \quad i = 1...n \qquad (C3)$$

Substituting (C3) into (C2),

$$\Delta P_T^H = \sum_{i=1}^{n} \lambda^i \Delta(CPI^i e^i) \tag{C4}$$

Similarly, movements in the CPI in the home country, ΔCPI^H, can be broken down into the changes in its traded and nontraded goods components:

$$\Delta CPI^H = a \Delta P_{NT}^H + (1 - a) \Delta P_T^H \tag{C5}$$

A standard formula for the real effective exchange rate of the home country is:

$$\Delta R = \Delta CPI^H - \sum_{i=1}^{n} \lambda^i \Delta(CPI^i e^i) \tag{C6}$$

where DR is the percentage change in the real effective exchange rate index.
Substituting (C4) and (C5) into (C6):

$$\Delta R = a(\Delta P_{NT}^H - \Delta P_T^H) \tag{C7}$$

It follows that, given the assumptions that the real exchange rates of the home country's trading partners are not changing (or that any such changes exactly cancel), and that the law of one price holds for traded goods, then movements in the conventionally measured real exchange rate index will be proportional to movements in the relative price of traded and nontraded goods.

NOTES

1. The author is indebted to Lindsay Hogan and Sally Thorpe for assistance in preparing this paper.

2. This analysis of a change in the terms of trade is somewhat similar to the analysis of a mineral 'boom' given by Snape (1977). There is, however, a subtle difference. In the case of new mineral discoveries or some other physical progress in the traded goods sector, the production possibilities frontier actually moves. The new production point can then be determined as the point of tangency between the new frontier and a relative price line. In contrast, in the case of a terms of trade change, the production point must still be determined on the actual rather than the "shadow" frontier.

REFERENCES

Barro, R.J. (1974). "Are Government Bonds Net Wealth?" Journal of Political Economy 82(6): 1095-117.

_____ (1979). "Determination of Public Debt." Journal of Political Economy 87(5): 940-71.

Bhandari, J.W. (1981). "Exchange Rate Overshooting Revisited." Manchester School of Economic and Social Studies 49(2): 165-72.

Branson, W.H. (1986). "The Limits of Monetary Coordination as Exchange Rate Policy." Brookings Papers on Economic Activity 1: 175-94.

Chalfant, J.A., G.C. Rausser, H.A. Love, and K.G. Stamoulis (1986). "The Effects of Monetary Policy on US Agriculture." Paper presented at the 30th Annual Meeting of the Australian Agricultural Economics Society, Canberra, February, 3-5.

Crowley, P.T., L.P. O'Mara, and R. Campbell (1983). "Import Quotas, Resource Development and Intersectoral Adjustment." Australian Economic Papers 22(41): 384-410.

Devadoss, S. and W.H. Meyers, (1987). "Relative Prices and Money: Further Results for the United States." American Journal of Agricultural Economics 69(4): 838-42.

Dixon, P.B., A.A. Powell, J. Sutton, and D. Vincent, (1982). ORANI: A Multi-Sectoral Model of the Australian Economy, Amsterdam: North-Holland.

Dixon, P.B. and B.R. Parmenter (1987). "Australia's Real Exchange Rate: 1985 to 1990." Paper prepared by Institute of Applied Economic and Social Research, University of Melbourne, for a public seminar on The Debt Crisis, Sydney, May 15.

Dornbusch, R. (1976). "Expectations and Exchange Rate Dynamics." Journal of Political Economy 84(6): 1161-76.

_____ (1986). "Flexible Exchange Rates and Excess Capital Mobility." Brookings Papers on Economic Activity 1: 209-35.

_____(1987). "The Dollar: How Much Further Depreciation Do We Need?" Federal Reserve Bank of Atlanta Economic Review (September/October): 2-13.

Dutton, J. and T. Grennes, (1985). Measurement of Effective Exchange Rates Appropriate for Agricultural Trade, Economics Research Report No. 51, Department of Economics and Business, North Carolina State University, Raleigh.

_____ (1987). "Alternative Measures of Effective Exchange Rates for Agricultural Trade." European Review of Agricultural Economics 14(4): 427-42.

Dwyer, J. (1987). "Real Effective Exchange Rates as Indicators of Competitiveness." BAE paper presented to the Economic Society of Australia, 16th Conference of Economists, Surfers Paradise, August 23-27.

Dwyer, J. and L. P. O'Mara (1988). "Measuring Australia's Competitiveness." Quarterly Review of the Rural Economy 10(1): 54-59.

Economic Planning Advisory Council (EPAC) (1986). External Balance and Economic Growth, Council Paper No. 22, Canberra: Office of EPAC.

_____ (1988). Australia's Medium Term Growth Potential, Council Paper No. 30, Canberra: Office of EPAC.

Economist (1988). "Trying to Hit a Moving Target." The Economist 27 (August): 57.

Fagerberg, J. (1988). "International Competitiveness." Economic Journal 98 (June): 355-74.

Frankel, J.A. and K.A. Froot (1986). "Understanding the US Dollar in the Eighties: the Expectations of Chartists and Fundamentalists." Economic Record 62(September Supplement): 24-38.

_____ (1988). "Explaining the Demand for Dollars: International Rates of Return, and the Expectations of Chartists and Fundamentalists," in P.L. Paarlberg and R.G. Chambers (eds), Macroeconomics, Agriculture and Exchange Rates, Boulder, Colorado: Westview Press, pp. 25-80.

Fraser, R. (ed.) (1987). Paying the Banker: Facing Australia's Foreign Debt Problem, Canberra: Australian Mining Industry Council.

Gray, M.R. and S.J. Turnovsky (1979). "The Stability of Exchange Rate Dynamics under Perfect Myopic Foresight." International Economic Review 20(3): 643-60.

Gregory, R.G. (1976). "Some Implications of the Growth of the Mineral Sector." Australian Journal of Agricultural Economics 20(2): 71-91.

Higgs, P.J. (1986). Adaptation and Survival in Australian Agriculture, Melbourne: Oxford University Press.

Hogan, L.I. (1986). "A Comparison of Alternative Exchange Rate Forecasting Models." Economic Record 62(177): 215-23.

Juttner, D.J. (1988). "Effective Exchange Rates." Economic Papers 7(2): 78-88.

Lewis, P. and J. Dwyer (1988). "Monetary Changes, Price Flexibility and Competitiveness." ABARE paper presented at the Australian Economic Congress, Economic Society of Australia, Canberra, August 28 - September 2.

Martin, W. and I. Shaw (1986). "The Effect of Exchange Rate Changes on the Value of Australia's Major Agricultural Exports." Economic Record 62(September Supplement): 101-7.

McKenzie, I.M. (1986). "Australia's Real Exchange Rate During the Twentieth Century." Economic Record 62(September Supplement): 69-78.

O'Mara, P. (1988). "The Medium Term Outlook for the Real Exchange Rate and Real Interest Rates." Review of Marketing and Agricultural Economics 51(1): 60-67.

O'Mara, P., D. Carland and R. Campbell (1980). "Exchange Rates and the Farm Sector." Quarterly Review of the Rural Economy 2(4): 357-67.

O'Mara, P., B. Crofts and R. Coote (1987). "Prospects for Exchange Rates and Interest Rates: Implications for Farmers." Quarterly Review of the Rural Economy 9(4): 440-8.

O'Mara, P., N.A. Wallace and H. Meshios (1987). "The Current Account, Monetary Policy, Market Sentiment and the Real Exchange Rate - Some Implications for the Farm Sector." Invited paper presented at the 31st Annual Conference of the Australian Agricultural Economics Society, Adelaide, February 10-12 [revised version forthcoming in Australian Journal of Agricultural Economics 31(3)].

Paarlberg, P.L. and R.G. Chambers (eds.) (1988). Macroeconomics, Agriculture and Exchange Rates, Boulder, Colorado: Westview Press.

Salter, W.E.G. (1959). "Internal and External Balance: the Role of Price and Expenditure Effects," Economic Record 35(71): 226-38.

Snape, R.H. (1977). "Effects of Mineral Development on the Economy." Australian Journal of Agricultural Economics 21(3): 147-56.

Stoeckel, A.B. (1979). "Some General Equilibrium Effects of Mining Growth on the Economy." Australian Journal of Agricultural Economics 23(1): 1-23.

Swan, T. (1955). "Longer Run Problems of the Balance of Payments." Melbourne: Paper presented to the Australian and New Zealand Association for the Advancement of Science. Published in H.W. Arndt and W.M. Corden (eds.) (1963). The Australian Economy: A Volume of Readings, Melbourne: Cheshire, pp.284-95.

_____ (1960). "Economic Control in a Dependent Economy." Economic Record 36(73): 51-66.

Thorpe, S., L. Hogan and R. Coote (1988). "Risk Premiums-- Implications for the Exchange Rate and Interest Rates in Australia since 1984." ABARE paper presented at a Department of Treasury seminar, Canberra: October 19.

Turnovsky, S.J. (1981). "Monetary Policy and Foreign Price Disturbances under Flexible Exchange Rates: a Stochastic Approach." Journal of Money, Credit and Banking 13(2): 156-76.

Wallace, N., K. Bramma and P. O'Mara (1988). "Current Account Imbalances of Major Western Economies: Causes, Policy Options and Some Implications for Australia." Quarterly Review of the Rural Economy 10(4): 392-402.

Williamson, J. (1985). The Exchange Rate System, Policy Analyses in International Economics No. 5 (September 1983, revised June 1985). Institute for International Economics, Washington DC.

6

Information, Expectations, and Foreign Exchange Market Efficiency[1]

Douglas K. Pearce

1. INTRODUCTION

Do asset prices largely reflect changes in "fundamentals" so that they serve as efficient aggregators of information and lead to efficient resource allocation? Until recently, the answer that most economists would have given to this question was yes. Asset price movements were seen as the appropriate responses to new information about relevant economic or political conditions by investors who displayed rational expectations. Recently, however, the paradigm of "efficient markets" has come under increasing attack. There is a resurgence of a Keynesian view of financial markets in which asset prices are substantially influenced by the moods of investors (or speculators). Numerous studies of stock returns report empirical anomalies to the efficient markets hypothesis such as excess returns in January, low returns on Mondays, and apparent overreactions to news.[2] In addition, influential papers by Shiller (1981, 1984) argue that stock prices are much too volatile to be compatible with the efficient markets model and that psychological influences play an important role. Naturally, the 1987 Crash did not hurt the credibility of this view.

Concern about the efficiency of the stock market has spread to the foreign exchange market. There is general consensus that exchange rates should be modeled in a manner similar to stock prices, as argued strongly by Mussa (1979) among others. From that perspective, Frenkel and Mussa (1980) and Bergstrand (1983) asserted that the volatility of flexible exchange rates was not a sign of inefficiency because it generally was less than the volatility of stock prices. This argument is vitiated if people no longer believe that stock prices are set according to the efficient markets model.

The question of the efficiency of these markets is, of course, of more than academic or even investing interest. Proponents of the inefficient markets view have made policy recommendations to reduce what they perceive as the harmful effects of stock price and exchange rate volatility. Reaching back to Keynes' General Theory, they have suggested that speculative excesses, the assumed source of much of the volatility, should be reduced by taxing transactions and imposing limits on price moves. More intervention by the Federal Reserve to reduce exchange rate variability has again been suggested, and some even advocate the Fed doing open market operations in equities [Fischer and Merton (1984)].

This controversy is familiar to macroeconomists since it is similar to the issue of whether the Fed should target interest rates or money growth.[3] If the economy is hit mostly with spending shocks, targeting the money supply is preferred because the spending shocks will be partially offset by changes in the interest rate in the same direction. The volatility in the interest rate is desirable, since the real economy would experience wider swings if the Fed targeted the interest rate. If, on the other hand, the shocks hitting the economy were due largely to an unstable money demand, stabilizing interest rates would insulate the real economy from the instability of the financial markets. In the first case the volatility of the interest rate is "good volatility" and in the second case it is "bad volatility." In the context of the exchange rate, if the market is efficient, then the volatility simply reflects the volatility of the fundamentals and intervention would add noise to the signal that the exchange rate conveys. In the "overshooting" model of Dornbusch (1976), for example, the exchange rate is volatile because it is compensating for a sluggish general price level.[4] There is no reason to offset this variability. If, however, the exchange rate movements are due to irrational expectations or speculative bubbles, intervention that reduces the volatility is desirable, assuming such volatility has real costs.[5]

This paper provides a selective review of recent research on the efficiency of the foreign exchange market, relating it to the coincident work on stock prices. Section 2 gives a brief description of the volatility of exchange rates over the last decade. Section 3 reviews the asset approach to modelling the exchange rate and compares it to the model for stock prices. Section 4 discusses tests of market efficiency and presents some results using recent data. Section 5 examines some explanations for the apparent violations of market efficiency in exchange markets. The final section offers some conclusions.

2. RECENT VOLATILITY IN EXCHANGE RATES

Changes in exchange rates attract a lot of attention these days, and there seems to be a perception that foreign exchange markets have become more volatile. For example, Paul Volcker (1988) recently asserted that "various measures of exchange-rate volatility--daily, monthly, and cyclically--have plainly increased." Chart 6.1 plots the weekly movements in the dollar exchange rates with the German

Chart 6.1

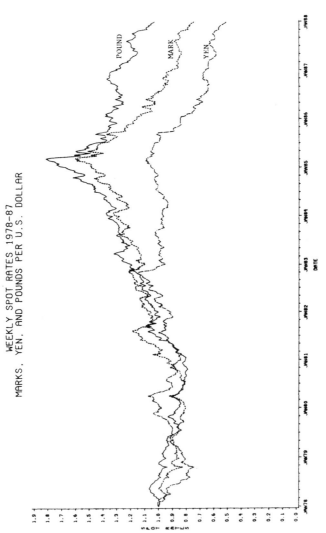

WEEKLY SPOT RATES 1978-87
MARKS, YEN, AND POUNDS PER U.S. DOLLAR

ALL SPOT RATES ARE NORMALIZED TO 1 ON JANUARY 6, 1978
SOURCE: FEDERAL RESERVE BOARD

mark, the British pound, and the Japanese Yen from 1978 to 1987, with exchange rates expressed as units of foreign currency per dollar. Each exchange rate has been normalized to one at the beginning of 1978. The chart illustrates that there has been considerable variation in rates over this period. The dollar rose against the mark and the pound until the beginning of 1985 and then fell back. The yen-dollar rate shows less variability, although the dollar also fell considerably in the last three years.

Has there been a change in the daily or weekly volatility in these exchange rates? Table 6.1 presents a measure of daily and weekly volatility for each year from 1978 to 1987. The top panel of the table gives the average daily absolute percentage change for each currency along with similar data for a portfolio of stocks, the Standard and Poors (S&P) 500. Two things are apparent from the table. First, stock prices generally display larger daily changes than any of the exchange rates. Second, there is no apparent secular change in daily volatility over this decade. The outlier for stocks is, not surprisingly, in 1987, but there is no corresponding increase in the volatility of exchange rates.

The lower panel presents similar data on week-to-week percentage changes. Again stock prices appear somewhat more volatile and there is no trend in volatility. As an alternative comparison, the lower panel also presents average absolute percentage changes in weekly cash prices for three commodities --wheat, cotton, and hogs--and for gold. Cash prices for these commodities were almost always more volatile than the exchange rates and usually more volatile than stock prices. While these descriptive statistics suggest that exchange rate volatility has not been rising, the issue of whether such volatility is excessive cannot be answered without a model that explains the sources of volatility.

3. FUNDAMENTAL MODELS OF EXCHANGE RATES

The notion of asset market efficiency requires a model for the determination of asset prices. The common characteristic of asset pricing models is that the price of the asset today is closely linked to the price expected to occur in the future. This in turn implies that the price of the asset today depends on expectations into the distant future of the underlying factors thought to influence the asset price. While this approach long has been used to model stock price determination, the asset approach to exchange rate modeling is more recent. The similarity of models, however, has led the literature to employ similar tests of the efficiency of each market. The approach also has been

TABLE 6.1. Volatility of Exchange Rates, Stock Prices, and Commodity Prices 1978-87

	1978	1979	1980	1981	1982	1983	1984	1985	1986	1987
A. Daily Absolute Percentage Change										
Mark	.492	.297	.426	.685	.467	.395	.556	.663	.568	.490
Yen	.537	.479	.534	.547	.593	.382	.312	.401	.522	.491
Pound	.362	.411	.375	.661	.460	.410	.501	.786	.511	.467
S&P 500	.607	.505	.824	.660	.851	.656	.608	.503	.669	1.135
B. Weekly Absolute Percentage Change										
Mark	1.180	.710	1.166	1.602	1.095	1.025	1.175	1.692	1.680	.972
Yen	1.366	.981	1.188	1.178	1.374	.952	.641	1.045	1.446	1.074
Pound	.870	1.106	.886	1.640	.996	1.056	1.133	1.872	1.140	1.052

Table 6.1 (continued)

	1978	1979	1980	1981	1982	1983	1984	1985	1986	1987
S&P 500	1.485	1.355	2.090	1.524	2.369	1.420	1.630	1.203	1.674	2.510
Wheat	2.241	2.428	2.421	1.854	2.059	1.024	1.451	1.874	2.715	2.363
Cotton	1.599	1.682	2.537	2.095	1.422	1.540	1.867	1.022	3.707	2.328
Hogs	2.686	2.636	3.101	2.624	2.422	3.059	2.381	2.604	2.924	3.003
Gold	- -	2.959	4.559	2.809	3.580	2.336	1.658	1.694	1.972	1.390

NOTES: S&P 500 price index is from the CRSP data tapes.

DATA SOURCES: The exchange rate and commodity data are from the Federal Reserve Board. Wheat is number 2 hard at Kansas City, cotton prices are an average of eleven markets, and hogs are Omaha prices.

used to model commodity prices and test the efficiency in these markets.[6]

For stock prices, the basic model assumes that the per share price today equals the present value of expected dividends:

$$SP_t = \sum_{i=1}^{\infty} d_t^i \, D^e_{t+i} \tag{1}$$

where SP_t = stock price

$d_t \quad = 1 \,/\, (1 + r_t)$

$D_t \quad$ = dividends per share

or that the expected or required rate of return, r_t , is

$$r_t = (\, D^e_{t+1} + (\, SP^e_{t+1} - SP_t \,)\,) \,/\, SP_t \tag{2}$$

Since expected dividends and the required rate of return are unobservable, direct tests of this model are impossible. Assumptions are required about the process generating dividends and the stability of the expected rate of return. In particular, if dividends are nonstationary, shocks to dividends would be considered permanent under rational expectations and would have a large impact on today's price. If the shocks were considered temporary, the effects should be much less. If it is assumed that expectations are rational and that the expected rate of return is a constant, it can be shown that actual returns should be random fluctuations about the constant expected return. Expected returns, however, need not be constant so that actual returns could be serially correlated without violating efficiency.

While the fundamental model of stock prices is noncontroversial as the benchmark model, the corresponding model for foreign exchange rates is a matter of considerable dispute. The model chosen depends on the assumptions made about asset substitution across countries and the flexibility of goods prices. The basic model, where s is the log of the spot exchange rate, is of the form:

$$s_t = z_t + \phi \, (s^e_{t+1} - s_t) \tag{3}$$

so that the spot rate today depends on the fundamentals today, z_t, plus the expected change in the spot rate. Iterating equation (3) ahead and applying rational expectations, implies that

$$s_t = \delta \; \Sigma \; \delta^i \; \phi^i \; z^e_{t+i} \quad \text{where} \quad \delta = (1/1+\phi) \text{ and } \quad i=0,\infty \quad (4)$$

The current spot rate should depend on the current value of fundamentals, z_t, and the "discounted" sum of all future expected values of the fundamentals.[7] As information arrives, agents are assumed to revise their expectations of the fundamentals and the spot rate adjusts to reflect these expectations. While equation (4) resembles equation (1), there are differences. Spot rates should depend on current fundamentals as well as on expectations of future conditions whereas stock prices depend only on expected future cash flows. This reflects the fact that foreign exchange has value as a medium of exchange as well as an asset. The discount factor for exchange rates depends on the parameter ϕ which, as illustrated below, depends on the model. In the stock market model the discount factor is assumed to be a function of the riskless rate of return plus a risk premium.

Specific models for the exchange rate impose different assumptions about the preferences of investors and the behavior of goods prices. Because a special case of the asset approach known as the monetary model is used in much of the literature, a simple version of that model will be employed to illustrate the asset approach. The monetary model assumes there are stable demand functions for money in each country (a two-country example will be used). The home country's money demand function is assumed to be

$$mh_t - ph_t = \alpha1 \; yh_t - \beta1 \; ih_t \qquad (5)$$

while the corresponding function for the foreign country is:

$$mf_t - pf_t = \alpha2 \; yf_t - \beta2 \; if_t \qquad (6)$$

where   mh, mf = home and foreign money supply in logs
        ph, pf = home and foreign price levels in logs
        yh, yf = home and foreign incomes in logs
        ih, if = home and foreign interest rates (not logs)

For simplicity, assume that the parameters are the same for each country and let m equal the difference in the countries' money supplies, foreign minus home (in logs), y equal the difference in the countries' incomes (in logs), p equal the difference in price levels (in logs), and i equal the interest differential (not in logs). Then

$$m_t - p_t = \alpha \; y_t - \beta \; i_t \qquad (7)$$

To derive the model for the spot exchange rate, we first assume that investors are risk neutral, and thus regard foreign and domestic securities as perfect substitutes. Under this assumption, the expected returns on investing in either country's bonds must be equal, that is, uncovered interest rate parity holds:

$$s^e_{t+1} - s_t = if_t - ih_t = i_t \qquad (8)$$

where again the exchange rate is defined as units of the foreign currency per unit of the home currency and s is the log of the exchange rate. Next, purchasing power parity is assumed:

$$s_t = pf_t - ph_t = p_t \qquad (9)$$

Substituting (8) and (9) into (7) and solving for s_t yields

$$s_t = r\, s^e_{t+1} + x_t \qquad (10)$$

where $r = \beta/(1+\beta)$ and $x_t = (1+\beta)^{-1} (m_t - \alpha\, y_t)$. Expanding this expression, the spot rate is

$$s_t = \sum_{i=0}^{\infty} r^i x^e_{t+i} \qquad (11)$$

Note that the discount factor depends on the semi-elasticity of money demand. In this model, a constant discount factor requires stable money demand functions.

If uncovered interest rate parity is relaxed, implying that investors are risk averse and, therefore, that domestic and foreign bonds are imperfect substitutes and purchasing power parity is also relaxed, we will still get a model that implies that the current spot rate depends on all the expected future values of the fundamental factors, but the set of factors changes. An error term for each money demand equation could also be included in z_t. This uncertainty about the "fundamental model" of exchange rates complicates tests of foreign exchange market efficiency.[8]

What does the asset approach imply for the time series behavior of exchange rates, and in particular, should we expect exchange rates to follow a random walk? If the asset model approach, equation (4), is correct, then the exchange rate would only follow a random walk if the fundamentals follow a random walk [see Hakkio (1986)]. In monetary models, this seems unlikely since the money supply does not follow a random walk. If changes in the fundamentals are difficult to predict, then we would expect exchange rates to be difficult to predict. Under the assumption that the forward rate is a rational expectation of the future spot rate, the forward rate should be

$$f_t = \mathop{E}_{t} s_{t+1} = \delta \Sigma \, \beta^i \delta^i \mathop{E}_{t} z_{t+i} \quad i=1,\infty \quad (12)$$

Since the spot rate and the forward rate depend on all the expected future values of the fundamentals, any shock that is viewed as persistent should cause the spot and forward rates to move closely together.

4. TESTS OF EFFICIENCY

Tests of the efficiency of asset markets comprise a vast literature. Most tests are of an indirect nature in that they test the implications of the efficient markets approach rather than testing whether actual asset prices conform to a particular model that the investigator believes is correct. Following Fama (1970), markets are said to be weak-form efficient if the past history of the asset price cannot be used to form a trading strategy that produces abnormal profits. Semistrong-form efficiency requires that any publically available information be already reflected in the asset price so that such information also cannot be used to generate abnormal profits. Finally, strong-form efficiency widens the information set to include private information. As Grossman and Stiglitz (1980) have argued, the notion that asset prices reflect all information is troublesome, since there must be some return to gathering costly information. Otherwise no one has an incentive to collect information and hence asset prices cannot reflect information. Thus, efficiency is usually taken to mean that the returns to using costly information are just enough for investors to earn a normal rate of return on their information expenditures.

The essence of market efficiency is that there should be no available information on which to base a trading strategy that yields abnormal profits when transactions costs are taken into account. This, in turn, requires a model of normal returns. For the stock market, the Capital Asset Pricing Model is usually employed to generate normal returns for individual stocks.[9] In the exchange rate literature, assumptions about the level of risk are required. If a trading strategy produces positive returns, are they large enough to cover transactions costs and to compensate for any risk exposure? The latter question is made difficult because, as Levich (1985) notes, "there is no agreement on the fundamental nature of foreign exchange risk, no adequate measure of it, and no model that determines the equilibrium fair return for bearing it" (p. 1024).

As mentioned in the introduction, there is considerable debate over whether the stock market is efficient. Even weak-form efficiency has been challenged, and many trading rules based on publically available information, have been asserted to generate excess returns.[10] This section reviews some of the tests that have been used to test efficiency in the foreign exchange markets and also discusses the broader tests of efficiency that examine whether our fundamental models can explain exchange rate movements.

4.1. Does the Exchange Rate Follow a Random Walk?

Mussa (1979) asserted that the log of the exchange rate should be well approximated by a random walk, consistent with weak-form efficiency. Under the asset model of exchange rates, we would expect exchange rates to follow a random walk only if the fundamental variables followed a random walk.[11] Several studies have examined whether exchange rates appear to follow a random walk, with differing results.[12] One complication is the possible intervention by central banks to smooth short-run movements in rates. This might cause shocks to be spread out over several periods and produce serially correlated changes.

To test for a random walk, the first step is to test whether exchange rates appear to have a unit root, that is, whether they are stationary. One approach is to estimate the following model: $[s = \log(\text{exchange rate})]$

$$Ds_t = s_t - s_{t-1} = a + b\ s_{t-1} + \sum_{i=1}^{n} c_i\ D\ s_{t-i} + u_t \qquad (13)$$

where u_t = serially uncorrelated error term with $E(u_t) = 0$.

The hypothesis is that the exchange rate is nonstationary, i.e. has a unit root, so that $a = b = 0$. The lagged values of the dependent variable are included to ensure serially uncorrelated errors. Table 6.2 reports the F-tests of this joint hypothesis for daily, weekly, and monthly data on the mark, yen and pound, all in logs. Critical values for this hypothesis have been computed by Dickey and Fuller (1981). In no case can the null hypothesis be rejected, and thus we conclude that all the exchange rate series are nonstationary.

The second stage of the random walk tests is to examine the estimated autocorrelation coefficients of the first differences of the logs of the exchange rates. Table 6.3 reports the largest estimated autocorrelation coefficient and the Box-Ljung statistic, which has a x^2 distribution under the null hypothesis that all autocorrelation coefficients are zero. For daily data, the estimated autocorrelation coefficients are small, but one can reject the hypothesis that all coefficients are zero for the yen and the pound. For weekly data, only the yen appears inconsistent with a random walk. For monthly data, one cannot reject the hypothesis that each exchange rate follows a random walk. These results suggest that there is some autocorrelation in changes in exchange rates for short intervals, possibly reflecting central bank intervention, but that monthly changes in exchange rates appear serially uncorrelated.[13]

An alternative approach to testing for weak-form efficiency is to see if filter rules produce abnormal profits. The idea of a filter rule is to base buying and selling decisions on whether the exchange rate is trending down or up and therefore is predicated on the assumption that there are persistent patterns. A peak is identified if the exchange rate has fallen by x percent from its recent high, while a trough is identified if the exchange rate has risen by x percent from its recent low. It would not be surprising to find that by trying all values of x, you would find at least one that produced high returns in any historical time series. Dooley and Shafer (1983), however, report that filter rules applied to daily data, which they had found profitable in earlier published work, continued to be profitable in a subsequent period even when transactions costs were accounted for. Sweeney (1986) also finds evidence of profitable filter rules. While this implies weak-form inefficiency if expected returns are constant, it is possible, as Sweeney notes, that the filter rule picks up the pattern of time-varying risk premia so that higher returns are simply compensating for higher risks.

4.2. Is the Forward Exchange Rate an Unbiased Prediction of the Future Spot Rate?

Much of the literature on the efficiency of the foreign exchange market centers on tests of whether the forward rate is an unbiased and efficient forecast of the future spot rate. If it is, then agents can use the forward rate as the best guide to future spot rates, producing the gains in resource allocation efficiency discussed by Stein (1981).

TABLE 6.2. Tests of Nonstationarity of Spot Exchange Rates
Data: January 1978-September 1988

Model: $D s_t = a + b s_{t-1} + R c_i D s_{t-i} + u_t$

Hypothesis: $a = b = 0$

Date Type	F Statistic[a]	Approximate Critical Value[b]
Daily data		
Mark	.336(2,2494)	4.59
Yen	2.140(2,2494)	4.59
Pound	.458(2,2494)	4.59
Weekly data		
Mark	.297(2,509)	4.61
Yen	1.207(2,509)	4.61
Pound	.513(2,509)	4.61
Monthly data		
Mark	.357(2,113)	4.71
Yen	.839(2,113)	4.71
Pound	.734(2,113)	4.71

NOTES: [a]Degrees of freedom in parentheses.
[b]Critical values from Table IV, Dickey and Fuller(1981).

DATA SOURCES: Harris Bank and the Federal Reserve Board.

TABLE 6.3. Random Walk Tests
          Data: January 1978-September 1988

Data Type	Largest Autocorrelation Coefficient	Box-Ljung Statistic
Daily		
Mark	.052 at lag 20	35.22 (24)*
Yen	.078 at lag 9	64.63** (24)
Pound	-.063 at lag 10	47.14** (24)
Weekly		
Mark	-.080 at lag 5	13.46 (12)
Yen	.154 at lag 2	25.73** (12)
Pound	.094 at lag 7	12.61 (12)
Monthly data		
Mark	.232 at lag 2	9.49 (6)
Yen	.136 at lag 4	4.34 (6)
Pound	.194 at lag 2	6.55 (6)

NOTES: * Degrees of freedom for Box-Ljung statistic,
          distributed as Chi-square
        ** Significant at the .05 level, indicating series is
          not white noise

DATA SOURCES: Harris Bank and the Federal Reserve Board.

Three assumptions lead to the prediction that the forward rate is an unbiased and efficient predictor of the future spot rate. First, we assume that there are no arbitrage profits available, which implies covered interest rate parity. Recalling that if_t is the foreign interest rate and ih_t is the domestic interest rate, covered interest rate parity is defined as:

$$if_t - ih_t = f_t - s_t \qquad (14)$$

where f_t is the log of the forward exchange rate and s_t is the log of the spot exchange rate, in units of foreign currency per unit of domestic currency. If (14) did not hold, there would be riskless profits available if transactions costs are zero. The second assumption is that investors are risk neutral so that uncovered interest rate parity holds:

$$if_t - ih_t = s^e_{t+1} - s_t \qquad (15)$$

That is, the interest rate differential equals the expected rate of appreciation of the domestic currency, assuming that the interest rates are defined over the appropriate time interval to make the units comparable. The third assumption is that expectations are rational

$$s^e_{t+1} = E(s_{t+1}|I_t) \qquad (16)$$

so that expectations are identical to the mathematical expectation of the future spot rate conditional on the available information, I_t. If all three assumptions hold, then the forward exchange rate must equal the rational expectation of the future spot rate

$$f_t = E(s_{t+1}|I_t) \qquad (17)$$

This in turn implies that the actual future spot rate should equal the forward rate plus a forecast error that is orthogonal to I_t.

To test the joint hypothesis of no arbitrage profits, risk neutrality and rational expectations, the following regression is often run:

$$s_{t+1} = \alpha + \beta f_t + \theta x_t + \epsilon_{t+1} \qquad (18)$$

The joint hypothesis is that $\alpha = 0$, $\beta = 1$, and $\theta = 0$, where x_t is any set of information known at time t. That is, the forward rate embodies all the information relevant for forecasting the spot rate so that any other information, x_t, is redundant, and hence the forward market is semistrong-form efficient. Often, x_t is omitted, since under the null its coefficient is zero, and thus its omission should not cause the simpler unbiasedness hypothesis, $\alpha = 0$ and $\beta = 1$, to be wrongly rejected. If risk aversion is present but the risk premium is constant, then we should find that $\alpha \neq 0$ but still find that $\beta = 1$ and $\theta = 0$.

The upper panel of Table 6.4 presents the results from tests of the simple unbiasedness hypothesis using nonoverlapping data on one-month ahead forward rates for the mark, yen and pound. The estimates indicate that the joint hypothesis of $\alpha = 0$ and $\beta = 1$ cannot be rejected. As Meese and Singleton (1982) point out, however, the test statistics assume that the series are stationary. The tests for unit roots discussed above indicate that this assumption is violated.

An alternative test of unbiasedness is to convert from levels to changes and estimate

$$s_{t+1} - s_t = \alpha + \beta[f_t - s_t] + e_t \qquad (19)$$

If the forward premium, $[f_t - s_t]$, is an unbiased estimate of the change in the spot rate, $\alpha = 0$ and $\beta = 1$ should still hold. The lower panel of Table 6.4 presents the estimates of this model. These results indicate that not only is the forward premium a biased forecast of the percentage change in the spot rate, the premium is significantly negatively related to the actual change. The low R^2s attest to how poor a predictor of change is the forward premium. Charts 6.2 through 6.4 illustrate how much more variable the actual change is compared to the forecasted change implied by the forward premia.

The evidence against the forward rate being an efficient predictor of the spot rate reported above is consistent with past work. Hansen and Hodrick (1980), Hakkio (1981), Baillie, Lippens, and McMahon (1983), and Hsieh (1984) employ weekly data, taking care to account for the autocorrelation introduced by using weekly observations on one-month forward rates, and report general rejection of the efficiency hypothesis. Geweke and Feige (1979), Bilson (1981), Longworth (1981), Huang (1984) and Gregory and McCurdy (1984) find similarly negative results using monthly data.[14]

The failure of the forward premium to be an unbiased predictor of changes in the spot rate means that the joint hypothesis of covered and uncovered interest rate parity and rational expectations must be rejected. Covered interest rate parity seems to have general empirical support [see Levich (1985, pp.1025-26) and Clinton (1988)]. Thus researchers have looked at whether risk premia exist under the assumption of rational expectations or investigated the rationality of survey measures of expectations.

TABLE 6.4. Tests of Unbiasedness of Forward Exchange Rates
As Predictors of Future Spot Rates

<u>Monthly data 1978(1) - 87(2)</u>

A. Levels $s_{t+1} = \alpha + \beta \ f_t + \epsilon_{t+1}$

	α	β	R^2	F	D.W.
Mark	.013 (.015)	.988 (.018)	.962	.764	1.71
Yen	.028 (.142)	.995 (.026)	.926	.062	1.59
Pound	-.002 (.009)	.993 (.015)	.974	.269	1.57

B. Changes $s_{t+1} - s_t = \alpha + \beta \ [f_t - s_t] + \epsilon_{t+1}$

	α	β	R^2	F	D.W.
Mark	-.014 (.007)	-3.649 (1.717)	.038	4.259*	1.86
Yen	-.015 (.005)	-2.696 (1.133)	.047	5.366*	1.80
Pound	.007 (.003)	-4.339 (1.079)	.126	12.516*	1.99

NOTES:All variables are in logs. The F statistic is for the
joint hypothesis that $\alpha = 0$ and $\beta = 1$. Standard errors of the
coefficients are in parentheses.

DATA SOURCES: Harris Bank and the Federal Reserve Board

Chart 6.2

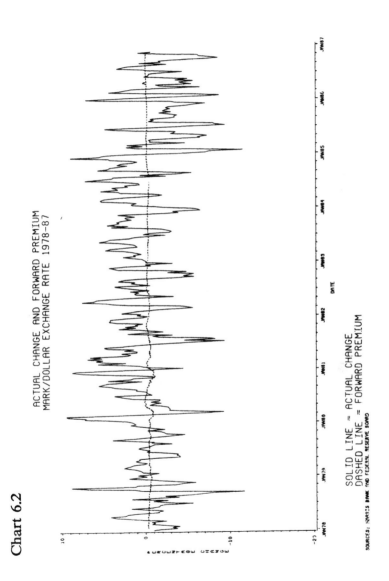

ACTUAL CHANGE AND FORWARD PREMIUM
MARK/DOLLAR EXCHANGE RATE 1978-87

SOLID LINE = ACTUAL CHANGE
DASHED LINE = FORWARD PREMIUM

SOURCES: HARRIS BANK AND FEDERAL RESERVE BOARD

Chart 6.3

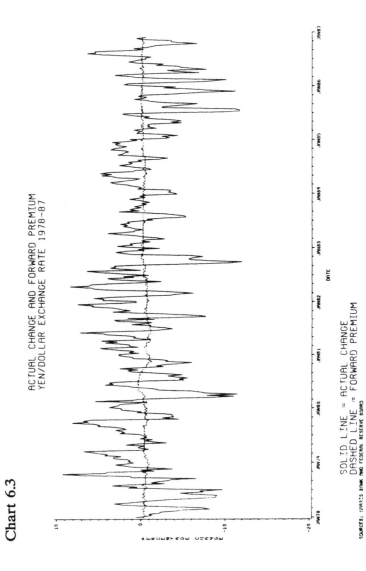

ACTUAL CHANGE AND FORWARD PREMIUM
YEN/DOLLAR EXCHANGE RATE 1978-87

SOLID LINE = ACTUAL CHANGE
DASHED LINE = FORWARD PREMIUM

SOURCES: NORRIS BANK AND FEDERAL RESERVE BOARD

Chart 6.4

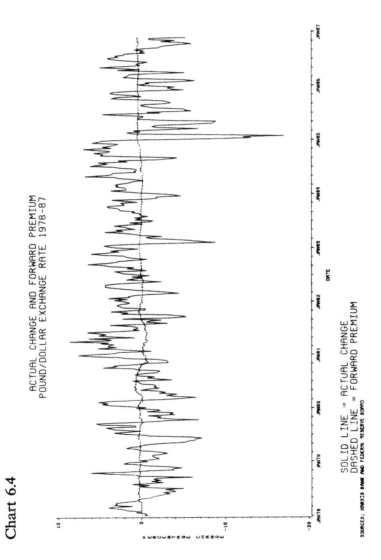

ACTUAL CHANGE AND FORWARD PREMIUM
POUND/DOLLAR EXCHANGE RATE 1978-87

SOLID LINE = ACTUAL CHANGE
DASHED LINE = FORWARD PREMIUM

SOURCES: MORRIS BANK AND FEDERAL RESERVE BOARD

If time-varying risk premia exist, then the forward rate is the sum of the expected future spot rate and a risk premium (rp):

$$f_t = s^e_{t+1} + rp_t \qquad (20)$$

Rational expectations implies that

$$s_{t+1} = s^e_{t+1} + u_{t+1} \qquad (21)$$

where u_{t+1} is a random error term. Substituting (20) into (21) yields

$$s_{t+1} = f_t - rp_t + u_{t+1} \qquad (22)$$

and subtracting s_t from both sides gives

$$s_{t+1} - s_t = f_t - s_t + (- rp_t + u_{t+1}) \qquad (23)$$

Equation (23) illustrates a problem introduced by time-varying risk premia, since the standard regression of the actual change on the forward premium would produce biased estimated parameters because the forward premium and the composite error term are correlated.[15] This problem is generally ignored.

Several studies have looked for time-varying risk premia under the maintained hypothesis of rational expectations. Fama (1984) examines the issue in the following way. He notes that if the current spot rate is subtracted from both sides of (20), we obtain

$$f_t - s_t = (s^e_{t+1} - s_t) + rp_t \qquad (24)$$

The forward premium is the sum of the expected change in the exchange rate and the risk premium. Fama estimates two complementary regression equations

$$f_t - s_{t+1} = a_0 + a_1 (f_t - s_t) + u_{t+1} \qquad (25)$$

$$s_{t+1} - s_t = b_0 + b_1 (f_t - s_t) - u_{t+1} \qquad (26)$$

The idea is that b_1 tells us if the forward premium can predict actual changes and a_1 tells us if the variation in the risk premium embedded in variations in the forward premium explains any of the deviation of the actual spot rate from the forward rate. Since $a_0 + b_0 = 0$ and $a_1 + b_1 = 1$, Fama hoped to divide the variation in the forward premium into variation due to expected changes and variation due to the risk premium. Consistent with the results reported above, however, he found that b_1 was negative. He concluded that if expectations are rational, then the variation in risk premia is much greater than the variation in expected changes, and that the risk premia must be negatively correlated with the expected change in the spot rate.[16]

While Fama found the latter result puzzling, Hodrick and Srivastava (1986), who replicated Fama's results, argue that this result is plausible. From (24) the risk premium is

$$rp_t = (f_t - s_t) - (s^e_{t+1} - s_t) \qquad (27)$$

Suppose the dollar is expected to depreciate by, say, 20 percent but the forward rate is only at a 10 percent discount from the current spot rate. The risk premium would be 10 percent, the expected profit from selling dollars forward for foreign exchange and using the proceeds to buy back dollars at the future spot rate. Hodrick and Srivastava show that it is not unreasonable to think this expected profit, the risk premium, should increase when the expected dollar depreciation is larger.[17]

The finding of time-varying risk premia means that the forward premium can be a biased forecast without implying inefficiency. This result would be more convincing if variations in the risk premia could be explained both theoretically and empirically. Several researchers have attempted to explain why the risk premium might vary over time. Frankel (1982) reports that relative asset supplies do not appear to determine risk premia. Hodrick and Srivastava (1984) examine the restrictions implied by Lucas's (1982) general equilibrium model and find no support for the model. Domowitz and Hakkio (1985) report evidence of a time-varying premium but tests of whether the premium was related to the conditional forecast error variance proved negative. Papers by Mark (1985) and by Cumby (1988) find little or no support for the intertemporal consumption model of risk premia derived from Breeden (1979). Bomhoff and Koedijk (1988) report more positive results that indicate that risk premia are affected by surprises in interest rates and inflation.

The papers examining risk premia take rational expectations as a maintained hypothesis. Recent work employing survey data on expected foreign exchange rates casts doubt on that assumption. Dominguez (1986) uses the median forecast from a survey of foreign exchange traders conducted by Money Market Services. For her limited data period of 1983 to 1985, she finds evidence that the forecasts were biased and hence inefficient. Frankel and Froot (1987) also examine the Money Market Services data along with several other surveys of exchange rate forecasts. They, too, report that the forecasts exhibit systematic bias, although they also find that the expectations appear inelastic and hence stabilizing. Avraham, Ungar and Zilberfarb (1987) conclude that depreciations were systematically underpredicted in Israel.

Has this bias in the Money Market Services data on exchange rate expectations been eliminated in recent years? Table 6.5 presents the results from regressing the actual spot rate on the expected spot rate and the actual change on the predicted change for the one-week and one-month ahead expectations. The survey data run from November 1984 through September 1988. As the upper panel of the table reports, using the level of the spot rate produces results that give mixed support for rationality, with the expectations for the mark and the yen appearing unbiased but those for the pound biased. Again, however, the test statistics may be misleading given the nonstationarity of the spot rates. The lower panel of the table reports the results for actual and predicted changes and these produce more negative results. All the weekly expected changes are biased and the monthly expected change is also biased for the pound. While the expected monthly changes for mark and yen are not unbiased in the sense that one cannot reject the joint hypothesis that the constant term is zero and the slope is one, the estimates for the yen are very imprecise. The low R^2s in all regressions for the change in exchange rates are consistent with the asset market approach in that it predicts that exchange rates will be difficult to predict.[18] Charts 6.5, 6.6, and 6.7 plot the actual and expected changes in the exchange rates computed from the monthly survey data and illustrate that most of the variation in exchange rates was unanticipated.

The evidence from the survey data suggests that the assumption of rational expectations may be unwarranted, and thus that the bias in the forward premium need not reflect risk premia. Of course all results based on such surveys can be challenged. The survey participants may not be representative or may not have an incentive to take the time to evaluate relevant information. In addition, as emphasized by Mishkin (1983), efficient asset markets do not require that all market participants have rational expectations.

4.3. How Do Exchange Rates React to News?

The asset market approach argues that most of the variation in spot exchange rates is due to revisions in expectations caused by new information. This implies that unpredicted changes in the exchange rate can be explained ex post by the news that occurred during the

TABLE 6.5. Tests of Unbiasedness of Survey Measures of Expected Exchange Rate Movements

A. Levels   Model: $s_{t+1} = a + b \ s^e_{t+1} + u_{t+1}$

Weekly	a	b	R^2	F(a=0;b=1)	D.W.
Mark	.028	.988	.994	2.278	1.95
	(.013)	(.006)			
Yen	1.455	.992	.997	1.923	1.88
	(.754)	(.004)			
Pound	.022	.967	.976	4.796*	1.91
	(.007)	(.011)			
Monthly					
Mark	.097	.955	.965	1.219	2.50
	(.062)	(.030)			
Yen	4.379	.968	.977	1.676	1.97
	(4.030)	(.022)			
Pound	.075	.880	.904	4.707*	1.74
	(.030)	(.043)			

B. Changes   Model: $s_{t+1} - s_t = a + b \ (s^e_{t+1} - s_t) + u_{t+1}$

Weekly					
Mark	-.084	.231	.008	9.260*	1.90
	(.128)	(.184)			
Yen	-.129	.361	.020	6.360*	1.90
	(.121)	(.181)			
Pound	-.036	-.562	.035	26.940*	2.15
	(.134)	(.213)			
Monthly					
Mark	.354	1.039	.103	.168	2.30
	(.507)	(.505)			
Yen	-1.005	.311	.008	1.289	1.94
	(.690)	(.522)			
Pound	-.423	-.340	.019	7.355*	1.95
	(.585)	(.372)			

NOTE: Standard errors in parentheses.

Chart 6.5

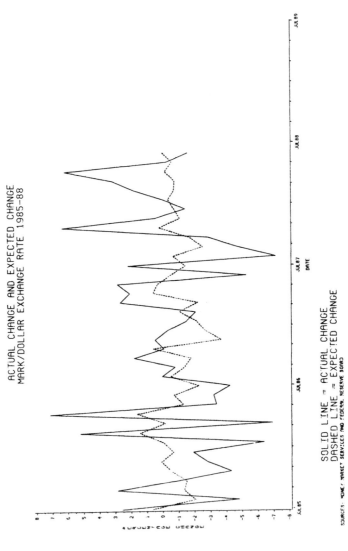

ACTUAL CHANGE AND EXPECTED CHANGE
MARK/DOLLAR EXCHANGE RATE 1985-88

SOLID LINE = ACTUAL CHANGE
DASHED LINE = EXPECTED CHANGE

SOURCES: MONEY MARKET SERVICES AND FEDERAL RESERVE BOARD

Chart 6.6

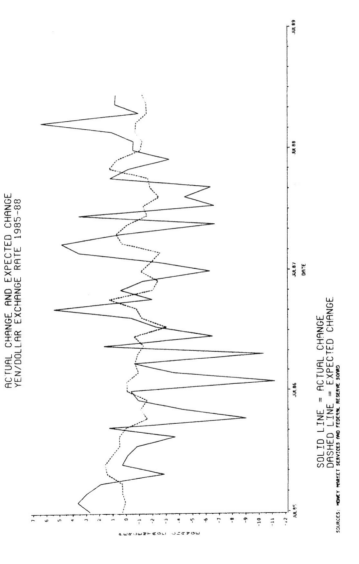

ACTUAL CHANGE AND EXPECTED CHANGE
YEN/DOLLAR EXCHANGE RATE 1985-88

SOLID LINE = ACTUAL CHANGE
DASHED LINE = EXPECTED CHANGE

SOURCES: MONEY MARKET SERVICES AND FEDERAL RESERVE BOARD

Chart 6.7

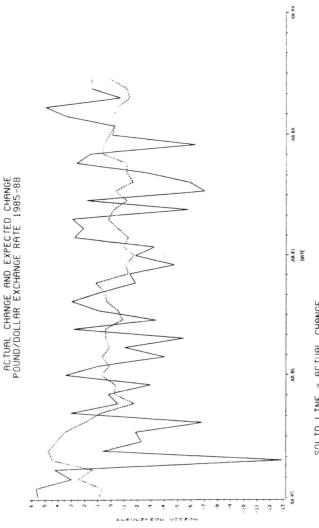

ACTUAL CHANGE AND EXPECTED CHANGE
POUND/DOLLAR EXCHANGE RATE 1985-88

SOLID LINE = ACTUAL CHANGE
DASHED LINE = EXPECTED CHANGE

SOURCES: MONEY MARKET SERVICES AND FEDERAL RESERVE BOARD

period. Dornbusch (1980) reports support for this view in that forecast errors for current accounts, economic growth rates, and interest rate differentials explained much of the unexpected change in exchange rates using the assumption of uncovered interest rate parity to generate the expected changes in exchange rates. His sample period was, however, quite short, consisting of semi-annual observations for only six years. Frenkel (1981) tests the "news" model by regressing the spot rate on the lagged forward rate, his measure of the expected spot rate, and on a measure of the unexpected interest rate differential, which he interprets as reflecting news on inflation. While Frenkel finds that the news variable is related to the unexpected movement in spot rates, the nonstationarity of the spot rate makes the statistical inferences suspect. Edwards (1982) employs a similar framework but includes news on money, real income, and real interest rates. He finds that countries with unexpectedly higher relative money growth experience depreciations.

Ott and Veugelers (1986) find that the significance of news in explaining ex post changes in exchange rates depends on the monetary regime. Only when they examined separately the October 1979-October 1982 period of money targeting did they find evidence that news mattered. During that period, their results suggest that news on expected inflation differentials was interpreted as a signal of subsequently tighter monetary policy and, thus, was associated with an appreciation of the dollar. Finally, Wolff (1988), using a VAR model to generate news on economic variables, finds no evidence that such news accounts for the observed exchange rate changes. Thus the ex post news approach to evaluating exchange market efficiency gives only mixed support.

An alternative approach to evaluating how news affects exchange rates is to examine how exchange rates move right after economic announcements. In an efficient market news that affects the expectations of fundamentals should be quickly reflected in asset prices. This implication has been investigated for foreign exchange rates by several researchers using an event study methodology much used in the finance literature.[19] For example, Hakkio and Pearce (1985) look at how spot exchange rates reacted to news about monetary policy and other economic variables. Observed changes in spot rates from just before to just after economic announcements are regressed on the expected and unexpected components of the announcements. If markets are efficient, prices should respond only to the unexpected components, the "surprises." They find that exchange rates rose immediately after an announcement that money growth had been higher than anticipated during the period in which the Federal Reserve was supposedly targeting money growth, October 1979 to October 1982, but not during the prior period when they were targeting interest rates. The response appeared to be largely completed within twenty minutes of the announcement. The results also indicate that anticipated money growth had no effect. Because the money announcement reports money growth in the recent past, the reaction of exchange rates implies that if growth is higher than anticipated, where the anticipation reflects the market's beliefs about the Federal Reserve's target growth rate, the market expects the Federal Reserve to slow money growth and this leads to an immediate appreciation of the dollar.[20] Exchange rates did not appear to respond to measures of unexpected inflation or unexpected real activity.[21]

One puzzling finding of the Hakkio-Pearce study was that there was also a positive response to money surprises from October 1982 to March 1984, a result also found by Hardouvelis (1988). Because the Federal Reserve abandoned its focus on money growth, the surprises should have had less effect. This may have been due to the Federal Reserve's change in operating procedures not being fully believed or understood by market participants. Estimating the model for the period March 1984 to December 1986, however, produces estimates that indicate that the response to unexpected money growth had disappeared, consistent with the view that the Federal Reserve no longer worried about money growth.

A recent paper by Deravi, Gregorowicz, and Hegji (1988) also reports evidence that the impact of news may be quite different over time. They look at how exchange rates respond right after the monthly trade deficit announcements for the United States. They find that prior to 1985 there was no effect, but that since then announcements of unexpectedly high deficits are followed by an immediate fall in the dollar. That part of the deficit announcement that was expected had no effect. They interpret their results as implying that either market participants expect even higher deficits in the future when the deficit is unexpectedly high, or that they believe that the Federal Reserve will respond to deficits by adopting policies to lower the dollar.

The responses of exchange rates to announcements generally are consistent with the efficient markets view. On the other hand, even when the change in the exchange rate is measured over a very short time interval, the news measures account for a relatively small fraction of the actual change.

4.4. How Well Do Fundamentals Explain the Volatility of Exchange Rate Movements?

Shiller (1981) created a considerable controversy by asserting that the fundamental model of stock prices could not explain the historical volatility of stock prices. Shiller's argument runs as follows. Suppose the expected rate of return is a constant. Today's stock price is a discounted sum of predicted future dividends. Since optimal forecasts must vary less than the variable they are forecasting, actual stock prices (the optimal forecasts) should vary less than a perfect foresight price that is computed ex post using the actual series of dividends and the assumed constant rate of return. Shiller's perfect foresight price series, however, varied less than actual stock prices, contradicting the efficient markets model. Subsequent research [see, for example, Kleidon (1986)], has argued that Shiller's conclusion can be overturned if dividends are nonstationary or if the expected rate of return is allowed to vary. This is still very much an active research question.[22]

Several researchers have applied some version of Shiller's approach to exchange rates to see if the actual variation can be explained by fundamentals. The basic approach is to assume a particular model, and hence a set of fundamentals, and to treat the parameters of the model as known and constant. If the spot rate less the current fundamentals is the discounted sum of expected future fundamentals, from equation (4),

$$s_t - \delta z_t = \Sigma \delta^i \phi^i z^e_{t+i} \qquad i=1,\infty \qquad (28)$$

then the variance of the left-hand side should be less than the variance of the right-hand side when the actual values of the z_{t+i} are substituted for the expected values. Huang (1981) computes a version of this type of "variance bounds" test for a monetary model and concludes that the exchange rate is excessively volatile. Wadhwani (1987) reaches a similar conclusion for a model based on that of Dornbusch (1976). Both these papers employ models without any stochastic error terms in the underlying structural models. As West (1987) has shown in the context of the monetary model, inclusion of such disturbances is enough to reverse the finding of excess volatility. Moreover, Diba (1987) points out that Huang used an estimate of the semi-elasticity of the demand for money that was much too small for monthly data and that simply using a more reasonable value of this parameter overturned Huang's results.

Variance bounds tests seem even more problematic for exchange rates than for stock prices. There is more uncertainty both about the underlying fundamentals and the constancy of parameters over different policy regimes. Moreover, the relatively small sample sizes for exchange rate models mean that inferences from volatility tests are suspect [see Frankel and Meese (1987)].

5. ALTERNATIVE EXPLANATIONS FOR APPARENT MARKET INEFFICIENCIES

As the above review suggests, there is considerable evidence that exchange markets are inefficient. Since this is a discomforting finding, it is not surprising that several explanations have been forthcoming. Clearly, if expectations are not rational, we need look no further since asset prices cannot reflect all information if expectations ignore relevant information. In this section, two alternative, although similar, explanations for the results are briefly discussed.

5.1. Speculative Bubbles

The recent literature on speculative bubbles is concerned with particular specifications of such bubbles that do not imply agents could expect to make abnormal profits. These are referred to as "rational speculative bubbles" by Blanchard and Watson (1982). These bubbles are usually assumed to reflect the influence of unobservable, to the econometrician, variables that cause the price of an asset to deviate from its "fundamental value" [Flood and Garber (1980)].

Consider a monetary model similar to that in Meese (1986). Let b_t be the bubble and s_t^* be the "fundamental" value of the exchange rate (in logs) so that

$$s_t = s_t^* + b_t \qquad (29)$$

To preserve the assumption of no arbitrage profits under risk neutrality, uncovered interest rate parity still has to hold so that equation (10) still holds

$$s_t = x_t + r\, s_{t+1}^e \qquad (10)$$

$$= x_t + r\, (s_{t+1}^* + b_{t+1})^e \qquad (30)$$

Following Blanchard and Watson, assume that each period the bubble is expected to continue with probability π and to burst with probability $(1 - \pi)$. In addition, it must be assumed that the bubble follows a particular process, namely

$$b_t = \pi^{-1} \, r^{-1} \, b_{t-1} + u_t \quad \text{if the bubble exists}$$

$$= u_t \qquad \qquad \text{if there is no bubble} \qquad (31)$$

where $E(u_t \mid I_{t-1}) = 0$. In this special case, the bubble can occur, but there are no arbitrage profits, since

$$s_t = s^*_t + b_t = x_t + r \, s^{*e}_{t+1} + r \, b^e_{t+1} \qquad (32)$$

and

$$b^e_{t+1} = E(b_{t+1} \mid I_t) = \pi \, (\, \pi^{-1} \, r^{-1} \, b_t)$$
$$= r^{-1} \, b_t \qquad \qquad (33)$$

Intuitively, the bubble would produce a rising exchange rate, but people's expectations would, during the bubble, appear to underpredict systematically the exchange rate because their expectations would reflect the probability that the bubble will burst. Uncovered interest rate parity would hold even though the forward premium would turn out to be a biased predictor during the bubble. Such bubbles would also make the exchange rate appear more volatile than the fundamental model would predict.

The evidence on whether bubble-like behavior can be found in recent exchange rate movements is mixed. Evans (1986) examines the pound-dollar exchange rate and notes that bubbles should produce runs of excess returns, where excess returns are measured as the difference between the spot rate and the lagged forward rate. He finds evidence of runs of negative excess returns on pound-denominated assets over 1981-84 consistent with the presence of a speculative bubble but is careful to note that such a finding is also consistent with nonrational expectations or, as discussed below, policy regime changes. Meese (1986) employs the monetary model and does specification tests of the joint hypothesis that the model is correct, the stochastic processes generating the fundamentals are constant, and there are no bubbles. He rejects the joint hypothesis for the pound-dollar and mark-dollar exchange rates over the period 1973-82, concluding that bubbles may have been present. Woo (1987) also finds evidence of speculative bubbles in several exchange rates using a portfolio balance model in which bonds of different countries are not assumed to be perfect substitutes. One problem with his approach is that the periods in which bubbles are found were picked a priori by looking for periods of abnormal exchange rate behavior.[23]

5.2. Regime Changes and Learning

In the speculative bubble model discussed above, the bubble is an exogenous variable. An alternative view is to consider that market participants are uncertain about the stochastic processes generating the fundamentals. Models incorporating such uncertainty can also display the characteristics associated with bubbles. As agents learn from realized values of the variables, they change their probabilities of which process, which regime, is relevant. The econometrician examining the exchange rate process may conclude that the market is inefficient because he does not take into account the possibility of regime changes. One example of this is the "peso problem" discussed in Krasker (1980) in which agents may be worried about the possibility of a regime change entailing a large depreciation of the currency. Examples that examine the consequences of agents' uncertainties about the money supply process are given in Harris and Purvis (1982) and Stulz (1987). In these cases, the exchange rate may appear to overreact to money supply changes because agents wrongly interpret a temporary change as a signal of a permanent policy change.[24]

To illustrate this issue, consider a simplified version of the framework developed in Lewis (1988). Assume that the only fundamental variable is the differential between the rate of growth of the foreign country's money supply and that of the home country, m_t. Using the monetary model, the spot rate is

$$s_t = m_t + \alpha (E\, s_{t+1} - s_t) \qquad (34)$$

Suppose that there are two regimes, with $m_t = \Theta_0 + u_t$ under the first regime and $m_t = \Theta_1 + u_t$ under the second, where u_t is a random error term with mean zero. Let $\Theta_0 > 0$ and $\Theta_1 = 0$ and let agents' estimates of the probability, at time t, that monetary policy follows the first regime be $P_{0,t}$.

Solving for s_t in the usual way yields:

$$s_t = (1 - r)\, \Sigma\, r^j\, E\, m_{t+j} \qquad \text{where } r = \alpha/(1+\alpha) \text{ and } j=0,\infty \quad (35)$$

At time t,

$$E\, m_{t+j} = \Theta_0\, P_{0,t} \qquad\qquad j>0 \qquad (36)$$

Substituting (36) into (35) yields [25]

$$s_t = (1-\tau) \, m_t + \tau \, \theta_0 \, P_{0,t} \tag{37}$$

The spot rate is simply a weighted average of today's money growth rate differential and the expected future differential formed at time t. Leading (37) ahead one period gives

$$s_{t+1} = (1 - \tau) \, m_{t+1} + \tau \, \theta_0 \, P_{0,t+1} \tag{38}$$

and so

$$E \, s_{t+1} = \theta_0 \, P_{0,t} \tag{39}$$

Suppose that the regime shifts from the first to the second at time (t + 1). Thus, the difference between the actual and expected exchange rate is

$$s_{t+1} - E \, s_{t+1} = (1 - \tau) \, u_{t+1} + \theta_0 \, (\tau \, P_{0,t+1} - P_{0,t}) \tag{40}$$

If agents slowly adjust their guess of P_0, then they will exhibit forecast errors that are systematically negative. That is, they will continue to believe that money growth will be faster in the foreign country and so expect a steady appreciation of the domestic currency, but the actual appreciation will be less. Presumably, they would eventually learn that the new regime is in place, but during this learning interval, the forward rate would exhibit a systematic bias. Lewis also shows that this type of model can account for deviations of the exchange rate from the ex post fundamentals and for what would appear as excess volatility. [26]

6. CONCLUSIONS

This paper has reviewed some of the recent literature on the efficiency of the foreign exchange market. This review has shown that there is considerable disagreement on whether the evidence is inconsistent with efficiency, reflecting in part the lack of agreement on the "fundamental model" for exchange rates. Exchange rates do not appear to follow random walks over very short time intervals, but the random walk model seems to be a good approximation for monthly data. There is no strong theoretical case, however, that suggests that the exchange rate need follow a random walk. There is consistent evidence that the forward premium is a very poor predictor of future changes in spot rates, but that is to be expected if exchange rates are viewed as asset prices. What is not consistent is that the evidence indicates that the forward premium is a biased predictor, and, in fact, is negatively correlated with actual changes.

Research investigating the reasons for this bias has taken several tacks. One strain of work assumes rational expectations and attributes the bias to time-varying risk premia. This work has not as yet been successful in explaining the determinants of such premia. A second strain has challenged the rational expectations assumption by looking at survey measures of expected exchange rates. Assuming such survey data are representative, the evidence from these studies are not favorable to the rational expectations hypothesis.

Other approaches to testing for efficiency examine whether actual exchange rate changes can be explained ex post by unanticipated changes in fundamentals and have met with limited success. Event studies that look at the immediate responses of exchange rates to expected and unexpected economic announcements generally find that the responses are consistent with efficient markets, but that such surprises account for little of the total change. Finally, some researchers have used the volatility tests developed by Shiller for analyzing stock prices, and have found evidence of "excess volatility," although these tests are difficult to interpret for exchange rates.

What might account for the apparent inefficiency of the foreign exchange market? Besides nonrational expectations, the paper discussed two alternatives. The first was that speculative bubbles might occasionally occur. The second is that existing models ignore the uncertainty that agents may have about the parameters of the model. Both these explanations can be shown to produce the kind of evidence of market inefficiency that has been found. While there is some empirical support for the existence of bubbles, this approach looks less promising. For believers in market efficiency, more work needs to be done to see if the models incorporating agents' uncertainties and their learning processes can successfully redeem market efficiency.

NOTES

1. The author thanks Tom Grennes, John Kitchen, Karlyn Mitchell, and Pierre Van Peteghem for helpful comments.

2. See Pearce (1987) for a review of this literature.

3. This approach to evaluating alternative monetary targets begins with Poole (1970).

4. See Frankel (1986) for an application of this model to an economy with flexible commodity prices and sluggish goods prices. Frankel shows that commodity prices may also overshoot their long-run equilibrium levels after a money supply increase to compensate for the slow adjustment of other prices. Obstfeld (1986) gives a variation in which commodity prices may not overshoot.

5. Obstfeld (1985) argues that "there is little direct evidence that exchange rate volatility per se has had a harmful effect on the allocation of resources or macroeconomic performance" (p. 433).

6. See Labys and Granger (1970) for early work supporting commodity market efficiency. Rausser and Carter (1983) look at the soybean futures market and Pindyck and Rotemberg (1988) evaluate the efficiency of several commodity markets jointly; both papers report evidence of market inefficiency.

7. Meese (1986) discusses the required transversality condition for this solution.

8. A review of the literature on estimating alternative exchange rate models is beyond the scope of this paper, but the well-known results of Meese and Rogoff (1983) illustrate the uncertainty about the "true" structural model.

9. Anomalies to stock market efficiency are often rationalized by arguing that the CAPM is an inadequate model for expected returns.

10. Poterba and Summers (1987) and Fama and French [1988] both report evidence that there is some predictability to stock prices over long periods. This implies either stock market inefficiency or changes in expected returns.

11. Since the first difference in the log of the exchange rate is essentially the percentage change, the log of the exchange rate might follow a random walk with drift if, using the monetary model for example, there was a constant money growth differential between the two countries.

12. See Kim (1987) and references cited therein.

13. Kim (1987) reports similar results for daily, weekly, and monthly data.

14. Much work has also been done evaluating the forecasting characteristics of commodity futures prices. In a survey of the work up to 1981, Kamara (1982) concludes that futures prices are poor forecasts for nonstorable commodities, but the results are more mixed for storable commodities. Similar results are reported by Garcia, Hudson, and Waller (1988).

15. Frenkel (1981) mentions a similar problem.

16. Fama and French (1987(employ a similar approach to study commodity futures prices. The results differ substantially across commodities. For example, they find that for broilers, eggs, hogs, and oats, the basis--the current futures price less the current cash price--has considerable forecasting power, explaining 20 to 40 percent of the actual changes in cash prices. Precious metals futures, however, appear to behave very similarly to forward exchange rates, with the basis negatively related to the actual change. See also Kitchen and Denbaly (1987) for an investigation of whether conditions analogous to covered and uncovered interest rate parity hold for commodity futures.

17. A recent paper by Boyer and Adams (1988) offers an alternative explanation. Using the monetary model, they argue that the risk premium gets reflected in interest rate differentials and this produces the negative correlation found by Fama (1984).

18. There is no evidence, however, that lagged forecast errors or lagged actual changes are correlated with the forecast errors.

19. See, for example, Cornell (1982), Engel and Frankel (1984), and Hardouvelis (1988).

20. Frankel and Hardouvelis (1985) examine the responses of commodity prices to unexpected changes in the money supply. They find that commodity prices generally fell after an announcement of unexpectedly rapid money growth during the period that the Federal Reserve was supposedly targeting the money supply, consistent with the view that unexpected money growth led to expectations of higher real interest rates. Husted and Kitchen (1985) report that foreign interest rates and forward premia respond to unexpected money supply changes in a way consistent with covered interest rate parity, although they find some evidence of market inefficiency in that expected money supply changes also had significant effects.

21. Survey data from Money Market Services were used to separate expected and unexpected components of announcements. The references in footnote 18 also employ these data. Unlike the exchange rate expectations series discussed above, the survey expectations data on money growth and other economic announcements do not appear to violate the rational expectations hypothesis, see Pearce and Roley (1985).

22. See Merton (1985), West (1988), and the recent exchange between Shiller (1988) and Kleidon (1988).

23. See Blanchard and Watson (1982), Obstfeld (1985), Singleton (1987), and Diba and Grossman (1988) for discussions of the theoretical and econometric problems in testing for bubbles.

24. Flood (1981) also considers how changes in monetary policy regimes can affect exchange rate volatility.

25. This follows because $(1-r) \Sigma\, r^j = r$ for $j=1,\infty$ and because it is assumed that the probability of the first regime occurring in any future period is $P_0 t$.

26. Borenstein (1987) and Tabellini (1988) consider similar models. See also Friedman (1979) for an analysis of learning under rational expectations and how it may result in expectations that appear ex post to be irrational.

REFERENCES

Avraham, D., M. Ungar and B-Z. Zilberfarb (1987). "Are Foreign Exchange Forecasts Rational?" Economics Letters 24(3): 291-93.

Baillie, R.T., R.E. Lippens, and P.C. McMahon (1983). "Testing Rational Expectations and Efficiency in the Foreign Exchange Market." Econometrica (May): 553-63.

Bergstrand, J.H. (1983). "Is Exchange Rate Volatility 'Excessive'?" New England Economic Review, Federal Reserve Bank of Boston (September/October): 5-14.

Bilson, J.F.O. (1981). "The 'Speculative Efficiency' Hypothesis." Journal of Business (July): 435-51.

Blanchard, O. and M.W. Watson (1982). "Bubbles, Rational Expectations, and Financial Markets," in P. Wachtel (ed.), Crises in the Economic and Financial Structure, Lexington, MA: Lexington Books, pp. 295-315.

Bomhoff, E.J. and K. G. Koedijk (1988). "Bilateral Exchange Rates and Risk Premium." Journal of International Money and Finance (June): 205-20.

Boothe, P.and D. Longworth (1986). "Foreign Exchange Market Efficiency Tests: Implications of Recent Empirical Findings." Journal of International Money and Finance (June): 135-52.

Borensztein, E.R. (1987). "Alternative Hypotheses About the Excess Return on Dollar Assets, 1980-84." IMF Staff Papers (March): 29-59.

Boyer, R.S. and F.C. Adams (1988). "Forward Premia and Risk Premia in a Simple Model of Exchange Rate Determination." Journal of Money, Credit and Banking (November): 633-44.

Breeden, D.T. (1979). "An Intertemporal Asset Pricing Model with Stochastic Consumption and Investment Opportunities." Journal of Financial Economics (September): 265-96.

Clinton, K. (1988). "Transactions Costs and Covered Interest Arbitrage: Theory and Evidence." Journal of Political Economy (April): 358-70.

Cornell, B. (1982). "Money Supply Announcements, Interest Rates, and Foreign Exchange." Journal of International Money and Finance (August): 201-08.

Cumby, R.E. (1988). "Is it Risk? Explaining Deviations from Uncovered Interest Parity." Journal of Monetary Economics (September): 279-99.

Deravi, K., P. Gregorowicz and C.E. Hegji (1988). "Balance of Trade Announcements and Movements in Exchange Rates." Southern Economic Journal (October): 279-87.

Diba, B.T. (1987). "A Critique of Variance Bounds Tests for Monetary Exchange Rate Models." Journal of Money, Credit and Banking (February): 104-11.

Diba, B.T. and H.I. Grossman (1988). "Rational Inflationary Bubbles." Journal of Monetary Economics (January): 35-46.

Dickey, D.A. and W.A. Fuller (1981). "Likelihood Ratio Statistics for Autoregressive Time Series with a Unit Root." Econometrica (July): 1057-72.

Dominguez, K.M. (1986). "Are Foreign Exchange Forecasts Rational." Economics Letters 21(3): 277-81.

Domowitz, I. and C.S. Hakkio (1985). "Conditional Variance and the Risk Premium in the Foreign Exchange Market." Journal of International Economic (August): 47-66.

Dooley, M.P. and J. Shafer (1983). "Analysis of Short-Run Exchange Rate Behavior: March 1973 to November 1981," in D. Bigman and T. Taya (ed.), Exchange Rate and Trade Instability, Cambridge, MA: Ballenger, pp. 43-69.

Dornbusch, R. (1976). "Expectations and Exchange Rate Dynamics." Journal of Political Economy (December): 1161-76.

Dornbusch, R. (1980)."Exchange Rate Economics: Where Do We Stand?" Brookings Papers on Economic Activity 1:143-85.

Edwards, S. (1982). "Exchange Rates and 'News': A Multi-Currency Approach." Journal of International Money and Finance (December): 211-24.

Engel, C.and J. Frankel (1984). "Why Interest Rates React to Money Announcements: An Explanation from the Foreign Exchange Market." Journal of Monetary Economics (January): 31-9.

Evans, G.W. (1986). "A Test for Speculative Bubbles in the Sterling-Dollar Exchange Rate: 1981-84." American Economic Review (September): 621-36.

Fama, E.F. (1970). "Efficient Capital Markets: A Review of Theory and Empirical Work." Journal of Finance (May): 383-417.

_____ (1984). "Forward and Spot Exchange Rates." Journal of Monetary Economics (November): 319-38.

Fama, E.F. and K.R. French (1987). "Commodity Futures Prices: Some Evidence on Forecast Power, Premiums, and the Theory of Storage." Journal of Business, (January): 55-73.

_____ (1988). "Permanent and Temporary Components of Stock Prices." Journal of Political Economy (April): 246-73.

Fischer, S. and R.C. Merton (1984). "Macroeconomics and Finance: The Role of the Stock Market." Carnegie-Rochester Conference Series on Public Policy (21): 57-108.

Flood, R.P. (1981). "Explanations of Exchange-Rate Volatility and Other Empirical Regularities in Some Popular Models of the Foreign Exchange Market." Carnegie-Rochester Conference Series on Public Policy (15): 219-50.

Flood, R.P. and P.M. Garber (1980). "Market Fundamentals versus Price-Level Bubbles: The First Tests." Journal of Political Economy (August): 745-70.

Frankel, J.A. (1982). "In Search of the Exchange Rate Premium: A Six-Currency Test Assuming Mean-Variance Optimization." Journal of International Money and Finance (December): 255-74.

Frankel, J.A. (1986). "Expectations and Commoity Price Dynamics: The Overshooting Model." American Journal of Agricultural Economics (May): 344-48.

Frankel, J.A. and K.A. Froot (1987). "Using Survey Data to Test Standard Propositions Regarding Exchange Rate Expectations." American Economic Review (March): 133-53.

Frankel, J.A. and G.A. Hardouvelis (1985). "Commodity Prices, Money Surprises, and Fed Credibility." Journal of Money, Credit and Banking (November, pt. 1): 425-38.

Frankel, J.A. and R. Meese (1987). "Are Exchange Rates Excessively Volatile?" in S. Fischer (ed.), NBER Macroeconomics Annual 1987, Cambridge, MA: MIT Press, pp. 117-53

Frenkel, J.A. (1981). "Flexible Exchange Rates, Prices, and the Role of "News": Lessons From the 1970s." Journal of Political Economy (August): 665-705.

Frenkel, J.A. and M.L. Mussa (1980). "The Efficiency of Foreign Exchange Markets and Measures of Turbulence." American Economic Review (May): 374-81.

Friedman, B.M. (1979). "Optimal Expectations and the Extreme Information Assumptions of 'Rational Expectations' Macromodels." Journal of Monetary Economics (January): 23-41.

Garcia, P., M.A. Hudson and M.L. Waller (1988). "The Pricing Efficiency of Agricultural Futures Markets: An Analysis of Previous Research Results." Southern Journal of Agricultural Economics (July): 119-30.

Geweke, J. and E. Feige (1979). "Some Joint Tests of the Efficiency of Markets for Forward Foreign Exchange." Review of Economics and Statistics (August): 334-41.

Gregory, A.W. and T.H. McCurdy (1984). "Testing the Unbiasedness Hypothesis in the Foreign Exchange Market: A Specification Analysis." Journal of International Money and Finance (December): 357-68.

Grossman, S.J. and J.E. Stiglitz (1980). "On the Impossibility of Informationally Efficient Markets." American Economic Review (June): 398-408.

Hakkio, C.S. (1981). "Expectations and the Forward Exchange Rate." International Economic Review (October): 663-78.

_____ (1986). "Does the Exchange Rate Follow a Random Walk? A Monte Carlo Study of Four Tests for a Random Walk." Journal of International Money and Finance (June): 221-29.

Hakkio, C.S. and D.K. Pearce (1985). "The Reaction of Exchange Rates to Economic News." Economic Inquiry (October): 621-36.

Hansen, L.P. and R.J. Hodrick (1980). "Forward Exchange Rates as Optimal Predictors of Future Spot Rates." Journal of Political Economy (October): 829-53.

Hardouvelis, G.A. (1988). "Economic News, Exchange Rates and Interest Rates." Journal of International Money and Finance (March): 23-35.

Harris, R.G. and D.D. Purvis (1982). "Incomplete Information and the Equilibrium Determination of the Forward Exchange Rate." Journal of International Money and Finance (December): 241-53.

Hodrick, R.J. and S. Srivastava (1984). "An Investigation of Risk and Return in Forward Foreign Exchange Markets." Journal of International Money and Finance (April): 5-29.

_____ (1986). "The Covariation of Risk Premiums and Expected Future Spot Exchange Rates." Journal of International Money and Finance (March, supplement): 5-21.

Hsieh, D.A. (1984). "Tests of Rational Expectations and No Risk Premium in Forward Exchange Markets." Journal of International Economic (August): 173-84.

Huang, R.D. (1981). "The Monetary Approach to Exchange Rates in an Efficient Foreign Exchange Market: Tests Based on Volatility." Journal of Finance (March): 31-41.

_____ (1984). "Some Alternative Tests of Forward Exchange Rates as Predictors of Future Spot Rates." Journal of International Money and Finance (August): 153-67.

Husted, S. and J. Kitchen (1985). "Some Evidence on the International Transmission of U.S. Money Supply Announcement Effects." Journal of Money, Credit and Banking (November, pt. 1): 456-66.

Kamara, A. (1982). "Issues in Futures Markets: A Survey." Journal of Futures Markets (Fall): 261-94.

Kim, B.J.C. (1987). "Do the Foreign Exchange Rates Really Follow a Random Walk?" Economics Letters 23(3): 289-93.

Kitchen, J. and M. Denbaly (1987). "Arbitrage Conditions, Interest Rates, and Commodity Prices." Journal of Agricultural Economics Research (Spring): 3-11.

Kleidon, A.W. (1986). "Variance Bounds Tests and Stock Market Valuation Models." Journal of Political Economy (October): 953-1001.

_____ (1988). "The Probability of Gross Violations of a Present Value Variance Inequality: Reply." Journal of Political Economy (October): 1093-96.

Krasker, W.S. (1980). "The 'Peso Problem' in Testing the Efficiency of Forward Exchange Markets." Journal of Monetary Economics (April): 269-76.

Labys, W.C. and C.W.J. Granger (1970). Speculation, Hedging and Commodity Price Forecasts, Lexington, MA: D.C. Heath.

Levich, R.M. (1985). "Empirical Studies of Exchange Rates: Price Behavior, Rate Determination and Market Efficiency," in R.W. Jones and P.B. Kenen (eds.), Handbook of International Economics, Amsterdam: North-Holland, pp. 979-1040.

Lewis, K.K. (1988). "The Persistence of the 'Peso Problem' When Policy is Noisy." Journal of International Money and Finance (March): 5-21.

Longworth, D. ("Testing the Efficiency of the Canadian- U.S. Exchange Market under the Assumption of No Risk Premium," Journal of Finance, March 1981, pp.43-49.

Lucas, R.E. (1982). "Interest rates and Currency Prices in a Two-Country World." Journal of Monetary Economics (November): 335-60.

Mark, N.C. (1985). "On Time Varying Risk Premia in the Foreign Exchange Market: An Econometric Analysis." Journal of Monetary Economics (July): 3-18.

Meese, R.A. (1986). "Testing for Bubbles in Exchange Markets: A Case of Sparkling Rates?" Journal of Political Economy (April): 345-73.

Meese, R.A. and K.Rogoff (1983). "Empirical Exchange Rate Models of the Seventies: Do They Fit Out of Sample?" Journal of International Economics (February): 3-24.

Meese, R.A. and K.J. Singleton (1982). "On Unit Roots and the Empirical Modeling of Exchange Rates." Journal of Finance (September): 1029-35.

Merton, R.C. (1987). "On the Current State of the Stock Market Rationality Hypothesis," in R.Dornbusch, S. Fischer and J. Bossons (eds.), Macroeconomics and Finance, Cambridge, MA: MIT Press, pp. 93-124.

Mishkin, F.S. (1983). A Rational Expectations Approach to Macroeconomics, Chicago: University of Chicago Press.

Mussa, M.L. (1979). "Empirical Regularities in the Behavior of Exchange Rates and Theories of the Foreign Exchange Market." Carnegie-Rochester Conference Series on Public Policy (11): 9-58.

Obstfeld, M. (1985). "Floating Exchange Rates: Experience and Prospects." Brookings Papers on Economic Activity (2): 369-450.

_____ (1986). "Overshooting Agricultural Commodity Markets and Public Policy: Discussion." American Journal of Agricultural Economics (May): 420-21.

Ott, M. and P.T.W.M. Veugelers (1986). "Forward Exchange Rates in Efficient Markets: The Effects of News and Changes in Monetary Policy Regimes." Review, Federal Reserve Bank of St. Louis (June/July): 5-15.

Pearce, D.K. (1987). "Challenges to the Concept of Stock Market Efficiency." Economic Review, Federal Reserve Bank of Kansas City (September/October): 16-33.

Pearce, D.K. and V.V. Roley (1985). "Stock Prices and Economic News." Journal of Business (January): 49-67.

Pindyck, R.S. and J. Rotemberg (1988). "The Excess Co-Movement of Commodity Prices." NBER Working Paper No. 2671 (July).

Poole, W. (1970). "Optimal Choice of Monetary Policy Instruments in a Simple Stochastic Macro Model." Quarterly Journal of Economics (May): 197-216.

Poterba, J.M. and L.H. Summers (1987). "Mean Reversion in Stock Prices: Evidence and Implications." NBER Working Paper 2343 (August).

Rausser, G.C. and C. Carter (1983). "Futures Market Efficiency in the Soybean Complex." Review of Economics and Statistics (August): 469-78.

Shiller, R.J. (1981). "Do Stock Prices Move Too Much to be Justified by Subsequent Changes in Dividends?" American Economic Review (June): 421-36.

_____ (1984). "Stock Prices and Social Dynamics." Brookings Papers on Economic Activity (2): 457-510.

_____ (1988). "The Probability of Gross Violations of a Prexent Value Variance Inequality." Journal of Political Economy (October): 1089-92.

Singleton, K. (1987). "Speculation and the Volatility of Foreign Currency Exchange Rates." Carnegie-Rochester Conference Series on Public Policy (26): 9-56.

Stein, J.L. (1981). "Speculative Price: Economic Welfare and the Idiot of Chance." Review of Economics and Statistics (May): 223-32.

Stulz, R. (1987). "An Equilibrium Model of Exchange Rate Determination and Asset Pricing with Nontraded Goods and Imperfect Information." Journal of Political Economy (October): 1024-40.

Sweeney, R.J. (1986). "Beating the Foreign Exchange Market." Journal of Finance (March): 163-82.

Tabellini, G. (1988). "Learning and the Volatility of Exchange Rates." Journal of International Money and Finance (June): 243-50.

Volcker, P.A. (1988). "Don't Count on Floating Exchange Rates." Wall Street Journal, November 28, p. A14.

Wadhwani, S.B. (1987). "Are Exchange Rates 'Excessively' Voltile?" Journal of International Economics (May): 339-48.

West, K.D. (1987). "A Standard Monetary Model and the Variability of the Deutschmark-Dollar Exchange Rate." Journal of International Economics August (1987): 57-76.

_____ (1988). "Bubbles, Fads and Stock Price Volatility Tests: A Partial Evaluation." Journal of Finance (July): 639-56.

Wolff, C.C.P. (1988). "Exchange Rates, Innovations and Forecasting." Journal of Interantional Money and Finance (March): 49-61.

Woo, W.T. (1987). "Some Evidence of Speculative Bubbles in the Foreign Exchange Market." Journal of Money, Credit, and Banking (November): 499-514.

Comments by John Kitchen

"Information, Expectations and Foreign Exchange Market Efficiency"

by Douglas Pearce

1. INTRODUCTION

Douglas Pearce presented a good review of the literature and ideas of his topic, and he also provided updated results and interpretation. His paper will serve as a useful reference.

This discussion concentrates on providing a link between the information presented by Pearce and agricultural trade and commodity issues, and covers four areas. First, the issue of exchange rate volatility and agricultural trade is addressed. Second, evidence on the bias of the forward premium and exchange rate expectations is used to exmaine the issue of optimal hedging. Third, the application of "news" approaches in commodity markets is discussed and some evidence presented. Finally, the role of policy uncertainty in exchange rate volatility is briefly discussed.

2. EXCHANGE RATE VOLATILITY AND TRADE

The proposition that exchange rate volatility produces negative impacts on trade has theoretical and empirical support. Hooper and Kohlhagen (1978) presented a model that explicitly showed the negative effect of exchange rate risk on international trade. However, their empirical results (for a sample period that included both fixed and flexible exchange rate periods) found no support for the hypothesis that exchange rate volatility had a negative impact on aggregate manufacturing trade volume. Later empirical studies (having sample periods that cover more of the flexible exchange rate period) by Cushman (1983), Akhtar and Hilton (1984), and Kenen and Rodrik (1986) found statistical support for the negative impact of exchange rate volatility on international trade, although the impacts varied across countries.

Two recent studies are particularly relevant here. Maskus (1986) found that U.S. agricultural trade was reduced by 6 percent over the 1974-84 period as a result of exchange rate risk. Further, of the sectors examined, the reduction in agricultural trade was the largest percentage reduction. Anderson (1988) provided a theoretical and empirical analysis of the role of exchange rate risk in the U.S. soybean export market. Her results revealed negative and significant impacts of exchange rate variability on U.S. soybean exports to France, Japan, and Spain, with exports to Japan being the least sensitive to exchange rate volatility. She interpreted the lesser sensitivity to exchange rate risk in the case of Japan to be due in part to the Japanese access to, and ability to use, forward and futures markets.

3. OPTIMAL HEDGING

The uncertainty and risk that arise from exchange rate and commodity price volatility can be reduced by covering open positions through forward or futures markets. Several issues addressed by Pearce have implications for the question of optimal hedging.

Kawai and Zilcha (1986) provided a theoretical model to analyze risk-averse exporter and importer behavior under both commodity price and exchange rate uncertainty. Their model applies to exporters and importers who face a price denominated in a foreign currency (e.g., in application, a non-U.S. exporter or importer). Several of their findings are sensitive to the issues addressed by Pearce. In particular, they found that the existence of forward and futures markets could produce large benefits for traders, provided the bias in the forward and futures market prices was not too large. On the issue of optimal hedging, Kawai and Zilcha found that the level of hedge in foreign exchange and futures markets varied directly with the bias in forward and futures prices in those markets. Hence, the observation by Pearce that "...not only is the forward premium a biased forecast of the percentage change in the spot rate, the premium is significantly negatively related to the actual change" has implications for the issue of optimal hedging.

The bias in forward exchange markets does in fact appear to be quite large, whether we want to attribute the bias to a risk premium as in Fama (1984) or to biased expectations as in Frankel and Froot (1987). In fact, Frankel and Froot found that during the extended period of U.S. dollar appreciation in the early 1980s both the forward exchange rate and the expected future spot rate consistently underpredicted the subsequent spot exchange rate. If such effects were pervasive across currencies, then hedging strategies and outcomes would be affected.

Consider the bias of the forward rate during the period of dollar appreciation: the dollar appreciation was consistently larger than what was predicted by the forward premium. If a foreign exporter was continuously in the market and had covered his foreign exchange exposure through the forward exchange market, then on average the exporter was worse off than if no hedges had been made. Large potential speculative profits existed for the exporter. During periods of dollar depreciation (the late 70s and after February 1985), evidence suggests that the opposite situation existed: on average, the dollar depreciation exceeded that predicted by the forward premium. Over the period of dollar depreciation, the exporter would have realized the highest profits by hedging completely.[1]

The issue is complicated even further by the question of biased exchange rate expectations. Frankel and Froot observed that the forward premium bias is attributable to biased expections (as measured in surveys) but also that the expectation bias exceeded the forward premium bias on average. If the survey data were representative of the foreign exporter and importer expectations, then foreign exporters and importers generally would have taken the wrong positions. For example, during the strong dollar appreciation, both expectations and the forward rate underpredicted the appreciation and expectations more so than the forward rate. If the exporter's (biased) expectation was for a lower dollar value than that predicted by the forward rate, then the exporter would have covered his position. As discussed above, on average that would have been the wrong decision.

In considering the role of exchange rates in promoting efficient trade, such results are not very reassuring. Note, though, that the implication drawn above is with respect to average profits. The evidence presented by Pearce shows that the variability of actual exchange rate changes greatly exceeds the variability of both expected changes and forward premiums. So, forward markets can still aid in reducing the risk associated with exchange rate variability, particularly for traders who are not continuously in the market.[2]

4. APPLICATIONS TO COMMODITY PRICES

While Pearce's primary interest was on exchange rates, he also presented some information and noted several citations concerning similar applications to commodity prices. Techniques developed for use in analyzing exchange rate and stock price behavior often have also been used to examine commodity prices.[3]

There has not been much success in clearly identifying the underlying fundamentals for exchange rates. Edwards (1983) found that the news variables he employed often had incorrect signs, and Meese and Rogoff (1983) were unable to identify an exchange rate model that could outperform a random walk even when the ex post realizations of the fundamentals were used.

In recent empirical work [Kitchen (1988)] I used a simple extension of the Frenkel (1981) and Edwards (1983) exchange rate news approaches in an application to corn, soybean, and wheat prices. Consider specifications for commodity prices similar to those used by Pearce (Table 6.4) for exchange rates:

$$s_{t+1} = a + b \, f_t = e_t \qquad (1)$$

and

$$s_{t+1} - s_t = a + b \, (f_t - s_t) + e_t \qquad (2)$$

where s_t is the natural logarithm of the spot commodity price in period t, f_t is the natural logarithm of the commodity futures price in period t for delivery in period $t+1$, and s_{t+1} is the natural logarithm of the spot price in period $t+1$.

Table D6.1 presents the results for regressions of (1) and (2). The F-statistics for the levels results indicate that the joint (a,b) = (0,1) hypothesis can be rejected for soybeans and wheat, although the results are suspect because of nonstationarity of prices and the low Durbin-Watson statistics for corn and wheat. The changes in specification results show that neither the a nor the b coefficent is significantly different from zero, but also that the joint (a,b) = (0,1) hypothesis cannot be rejected. The Durbin-Watson statistics for corn and wheat are in the inconclusive range.

The specification in (2) can be extended to include news as in Frenkel (1981), where the news represents unexpected changes in the underlying fundamentals between t and $t+1$:

$$s_{t+1} - s_t = a + b \, (f_t - s_t) + news(t,t+1) + e_t \qquad (3)$$

Specifications based on (3) allow for more efficient estimation of parameters and also can provide information on the roles and importance of various news variables.

The news variables examined included interest rates, exchange rates, precipitation, temperature, exports, private stocks, and loan rates.[4] A slope dummy variable was used for interest rate news in an attempt to capture the effect of the 1979-82 monetary aggregate targeting experiment (Dum = 1 for October 1979 - October 1982, Dum = 0 otherwise).

Estimation results suggested that interest rate news (UDR) and weather news were very important in explaining price changes, while news about exchange rates, exports, private stocks, and loan rates did

TABLE D6.1. Commodity Price Regressions

A. Levels $s_{t+1} = a + b\, f_t + e_{t+1}$

	a	b	R^2	DW	F
Corn	.996	.814	.65	1.34	1.88
	(.607)	(.110)			
Soybeans	2.658	.583	.45	1.85	6.05*
	(.765)	(.120)			
Wheat	1.772	.693	.61	1.22	4.40*
	(.602)	(.104)			

B. Changes $s_{t+1} - s_t = a + b\,(f_t - s_t) + e_{t+1}$

	a	b	R^2	DW	F
Corn	-.022	.941	.066	1.45	0.48
	(.040	(.671)			
Soybeans	.008	.319	.012	2.19	1.53
	(.041)	(.552)			
Wheat	.018	.009	.001	1.46	2.35
	(.040)	(.458)			

NOTES: The sample period extends from 1971 through 1986. Prices were drawn based on a six-month sampling and forecasting horizon so that overlapping data problems would not be present. Number of observations is 32. The prices used were futures prices at close of the first business days of March and September for March and September contracts. F is the calculated value of the F-statistic for testing the joint (a,b) = (0,1) hypothesis. * indicates significance at the 5 percent level.

not enter significantly. The best news specification for corn was (standard errors in parentheses, F-statistic for testing hypothesis $(a,b) = (0,1)$):

$$s_{t+1} - s_t = \begin{array}{l} -.017 + 1.05 \ (f_t - s_t) + .073 \ \text{UDR} - .355 \ \text{DumUDR} \\ (.026) \ (.45) \qquad\qquad (.039) \qquad (.071) \end{array}$$

$$\begin{array}{l} + .109 \ \text{Aprrain} - .083 \ \text{Mayrain} + .116 \ \text{Junrain} \\ (.042) \qquad\qquad (.036) \qquad\qquad (.056) \end{array}$$

$$\begin{array}{l} + .083 \ \text{Seprain} - .032 \ \text{Maytemp} \\ (.030) \qquad\qquad (.013) \end{array} \quad -$$

$$\bar{R}^2 = .75 \qquad\qquad \text{DW} = 1.91 \qquad\qquad F = .477$$

For soybeans:

$$s_{t+1} - s_t = \begin{array}{l} -.001 + .81 \ (f_t - s_t) + .109 \ \text{UDR} - .263 \ \text{DumUDR} \\ (.033) \ (.46) \qquad\qquad (.056) \qquad (.102) \end{array}$$

$$\begin{array}{l} + .106 \ \text{Seprain} - .040 \ \text{Mayrain} \\ (.047) \qquad\qquad (.016) \end{array}$$

$$\bar{R}^2 = .45 \qquad\qquad \text{DW} = 1.70 \qquad\qquad F = .212$$

For wheat:

$$s_{t+1} - s_t = \begin{array}{l} .046 + .65 \ (f_t - s_t) + .182 \ \text{UDR} - .417 \ \text{DumUDR} \\ (.030) \ (.35) \qquad\qquad (.046) \qquad (.102) \end{array}$$

$$\begin{array}{l} - .127 \ \text{Mayrain} + .247 \ \text{Julrain} + .091 \ \text{Seprain} \\ (.046) \qquad\qquad (.065) \qquad\qquad (.037) \end{array}$$

$$\bar{R}^2 = .58 \qquad\qquad \text{DW} = 2.20 \qquad\qquad F = 1.86$$

A quick review of some of the effects: Excess rain during planting (May, corn) and harvest periods (July, wheat; September, corn, soybeans and wheat) resulted in higher prices. Commodity prices generally were positively related to interest rate shocks (interest rate changes dominated by inflation expectations). However, during the period when Federal Reserve policy was dominated by monetary aggregate targets, commodity prices were negatively related to unexpected interest rate changes (interest rate shocks were predominantly real interest rate shocks).

The improved efficiency of the estimation leads to results that are more in line with market efficiency. The a coefficient estimates are not significantly different from zero, while the b coeffcients are signficantly different from zero (.05 level for corn, .10 level for soybeans and wheat) but not significantly different from 1. The Durbin-Watson statistics suggest that there is no problem with serial correlation. Finally, as revealed by the F-statistics, the joint hypothesis that (a,b) = (0,1) cannot be rejected.

These results reveal a distinction between commodity markets and exchange markets--the fundamentals, and the effects that fundamentals have on the price variable, are somewhat easier to identify for commodity prices than for exchange rates. Also, similar to evidence presented in Fama (1984) and Fama and French (1987), commodity futures prices appear to have better forecasting power than do forward exchange rates. However, as evidence presented by Pearce shows, commodity prices are much more volatile than exchange rates, so there is a lot more variation that can be explained.[5]

5. THE BROADER QUESTION: SOURCE OF VOLATILITY

Failure to understand the determination of exchange rates in conjunction with the underlying uncertainty about economic relationships in general--and economic policies in particular--suggests that we won't see much change in the volatility of exchange rates. The U.S. experience of the past decade and a half--accommodative monetary policies followed by contractionary monetary policies and then large Federal budget deficits--suggests that policy uncertainty is persistent. The primary point made by Pearce is that the exchange rate is a highly flexible price that reacts to new information. The volatility in exchange rates is simply a reflection of underlying volatility and uncertainty, particularly volatilty and uncertainty about policies. As long as policy makers continue to create a policy environment that promotes uncertainty, the exchange rate will continue to act as a price that reflects that uncertainty.[6]

NOTES

1. The foreign importer would have had opposite opportunities. On average, the profitable strategies would have been to hedge completely during the dollar appreciation and to have complete exposure during the depreciations.

2. This view reveals some potential benefits that would promote "national marketing boards" or other methods of pooling exposure to risk.

3. Notable examples are the futures market forecasting accuracy and risk premium study by Fama and French (1987) and the overshooting model and evidence presented in Frankel (1986), and Frankel and Hardouvelis (1985).

4. Monthly rainfall and temperature values were obtained from Weather in U.S. Agriculture. For corn and soybeans, data for the Corn Belt farm production region were used. For wheat, data for the Northern and Southern Plains production regions were used. The news for each month was defined as the deviation from a 35-year moving average for the month. The news components for the other variables were determined using an instrumental variables approach employing Durbin's rank variable, an approach similar to that used by Frenkel. Details concerning the data and interpretation of empirical results are available in Kitchen (1988).

5. Also, in the specifications shown a large part of the variation remains unexplained, as in the case of soybeans where $R^2 = .45$.

6. See a paper by Willett (1986) for additional discussion along these lines.

REFERENCES

Akhtar, M.A. and R. Spence Hilton (1984). "Exchange Rate Uncertainty and International Trade: Some Conceptual Issues and New Estimates for Germany and the United States," Federal Reserve Bank of New York, Quarterly Review.

Anderson, Margot (1988). "U.S. Soybean Trade and Exchange Rate Volatility," U.S. Department of Agriculture, Economic Research Service, Technical Bulletin 1748, (October).

Cushman, David O. (1983). "The Effects of Real Exchange Rate Risk on International Trade." Journal of International Economics (August): 45-63.

Edwards, Sebastian (1983). "Floating Exchange Rates, Expectations, and New Information." Journal of Monetary Economics (May): 321-36.

Fama, Eugene F. (1984). "Forward and Spot Exchange Rates." Journal of Monetary Economics (November): 319-38.

Fama, Eugene F. and Kenneth R. French (1987). "Commodity Futures Prices: Some Evidence on Forecast Power, Premiums, and the Theory of Storage." Journal of Business (January): 55-73.

Frankel, Jeffrey A. (1986). "Expectations and Commodity Price Dynamics: The Overshooting Model." American Journal of Agricultural Economics (May): 344-48.

Frankel, Jeffrey A. and Kenneth A. Froot (1987). "Using Survey Data to Test Standard Propositions Regarding Exchange Rate Expectations." American Econmomic Review (March): 133-53.

Frankel, Jeffrey A. and Gikas A. Hardouvelis (1985). "Commodity Prices, Money Surprises and Fed Credibility." Journal of Money, Credit, and Banking (November): 425-37.

Frenkel, Jacob A. (1981). "Flexible Exchange Rates, Prices, and the Role of 'News': Lessons from the 1970s." Journal of Political Economy (August): 665-705.

Hooper, Peter and Steven W. Kohlhagen (1978). "The Effect of Exchange Rate Uncertainty on the Prices and Volume of International Trade." Journal of International Economics (November): 483-511.

Kawai, Masahiro and Itzhak Zilcha (1986). "International Trade with Forward-Futures Markets under Exchange Rate and Price Uncertainty." Journal of International Economics (February): 83-98.

Kenen, Peter B. and Dani Rodrik (1986). "Measuring and Analyzing the Effects of Short-term Volatility in Real Exchange Rates." Review of Economics and Statistics (May): 311-15.

Kitchen, John (1988). "Agricultural Futures Prices and New Information," U.S. Department of Agriculture, Economic Research Service, Staff Report AGES880805 (August).

Maskus, Keith E. (1986). "Exchange Rate Risk and U.S. Trade: A Sectoral Analysis." Federal Reserve Bank of Kansas City, Economic Review (March): 16-28.

Meese, R.A., and K. Rogoff (1983). "Empirical Exchange Rate Models of the Seventies: Do They Fit Out of Sample?" Journal of International Economics (February): 3-24.

Willett, Thomas D. (1986). "Exchange-Rate Volatility, International Trade, and Resource Allocation." Journal of International Money and Finance (March Supplement): S101-112.

7

Fiscal Policy, Exchange Rates, and World Debt Problems

Douglas D. Purvis

"Ronald Reagan made America walk tall partly because he placed it on a mountain of debt. Sooner or later, mountains have avalanches."
The Economist, September 10, 1988

1. INTRODUCTION

The topic on which I was invited to speak, "Fiscal Policy, Exchange Rates, and World Debt Problems," is of great interest to all observers of the international economy. Indeed, it has been of great interest for a number of years now, but policy progress has been slow; the term "muddling through" seems best to describe the "solutions" offered to date. Despite this, the topic takes on renewed importance at this time as a new president assumes office and once again debate about alternative plans to deal with the U.S. "twin deficits" takes center stage; it is deja vu--or as Time Magazine put it, deja voodoo economics--all over again.

The focus then is on the U.S. government budget deficit and the U.S. current account deficit; both have been at record--and in some quarters, alarming--levels for the past few years. In the first section I review some pertinent facts which describe the world economy's current situation and give some insights into how it evolved. I then turn to a theoretical discussion of the effects of fiscal policies in an open economy; the concern is not only with the traditional domestic stabilization issue but also with the international transmission of fiscal policies and their role in explaining the extreme volatility of exchange rates that has been experienced. Then, in Section 3, I review the prospects for an orderly unwinding of international imbalances and for the hard-landing alternative. I conclude with some remarks about proposals to reform the international monetary system.

2. THE EVOLUTION OF THE TWIN DEFICITS AND THE EXCHANGE RATE

The 1970s witnessed emergence of stagflation, driven primarily by sharp increases in the prices of oil and other primary commodities and foodstuffs. At the end of the decade, triggered by "OPEC II," the United States identified inflation as Public Enemy No. 1 and initiated a sharp monetary contraction. That contraction provoked a sharp recession and led to a dramatic fall in inflation from the double digit levels it had attained to the 4 percent range, where it now has remained for a number of years. This disinflation was accompanied by a sharp increase in interest rates and a sharp appreciation of the U.S. dollar; from the American perspective, the appreciation was welcome as it depressed import prices and reinforced the disinflation.

From the overseas perspective, the parallel depreciation served to exacerbate inflationary problems; the well-known beggar-thy-neighbor aspects of monetary policy under flexible exchange rates meant that the United States had succeeded in "exporting" its inflation. Hence, a number of overseas economies--also wanting to disinflate--chose to raise their interest rates to defend their currencies. Consequently, there occurred a substantial, concerted worldwide monetary contraction that led in turn to a massive worldwide recession. The degree of monetary contraction was much more severe than any of the central banks might have preferred "ex ante," although "ex post" some people believe a severe shock was required to break the inertia of inflation.[2]

The next move in this historical vignette went to the supply siders in the American administration who were successful in promoting massive tax cuts as a panacea for all that ailed the economy. These tax cuts remain controversial today; while they did not provide the promised panacea, they did induce the record government deficits that are still with us and they did contribute unwittingly to a Keynesian-style, fiscally led recovery from the recession. As a side effect, the interest and exchange rate effects of the earlier monetary contraction were sustained through the early years of the recovery. Further, these factors contributed to the emergence of the twin deficits and the accompanying inflows of foreign capital. In explaining the latter, the supply-siders stress the role of the tax cuts and improved profit potential in attracting capital, whereas conventional analysis stresses the trade deficit, which is attributed to the expenditure-switching effects of the high exchange rate and the expenditure-increasing effects of the government deficit.

Of course, the U.S. trade deficit had to be reflected in trade surpluses of its trading partners (except for the $100 billion or so errors and omissions item in the international reconciliation!), and these tendencies toward surplus were reinforced by fiscal contraction in a number of those countries. This "disharmonization" of fiscal policies thus contributed to the size and persistence of the external imbalances.

Figure 7.1 shows the extent of the external imbalances that emerged.[3] The first two years in the chart are representative of the size if not the particular pattern of the imbalances that had prevailed in previous years. The growth in the imbalances in subsequent years is thus all the more remarkable. The U.S. deficit on current account swelled to over $160 billion in 1987, about 3.6 percent of GNP; the overseas surpluses also reached record levels, with Japan being the largest but by no means the only significant surplus unit.

These external flows have a cumulative effect on the net foreign asset position of the nations involved. Most dramatic is the U.S. experience; having entered the decade as the world's largest creditor nation (with net foreign assets of $150 billion), the United States is now the world's largest net debtor nation. U.S. net foreign debt is now over $500 billion, and by early next decade it is projected to reach $1 trillion!

The evolution of the U.S. federal government deficit over the 1980s is shown in Figure 7.2. The surge in the fiscal deficit in the 1981-83 period reflects the combined effects of the cyclical reduction in tax revenues, discretionary tax cuts, and increased government expenditures. The deficit reached $220 billion (5.3 percent of GNP) in fiscal year 1986. On a general government basis (i.e., incorporating state and local governments), the deficit is smaller but still at record levels.

The relationship between the external deficits in Figure 7.1 and the government deficits in Figure 7.2 can best be viewed in terms of the national savings identities--Courchene quips that the global imbalances problem thus represents an "international identities crisis"!

The basic relation is derived from the open economy national income identity that national income, Y, equals domestic consumption, C, plus investment, I, plus government spending, G, plus the trade surplus, NX. This is shown in equation (1):

$$Y = C + I + G + NX \qquad (1)$$

From (1) it can be seen that the trade account deficit (NX < 0) reflects an excess of domestic expenditure over income; discussions

274

Figure 7.1

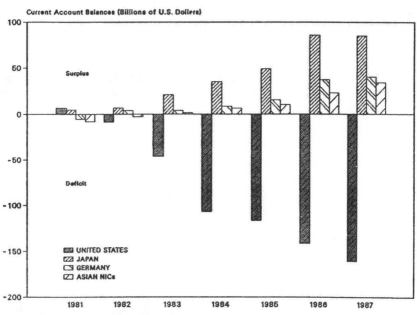

KEY INTERNATIONAL TRADE IMBALANCES IN THE 1980s

Current Account Balances (Billions of U.S. Dollars)

Surplus

Deficit

UNITED STATES
JAPAN
GERMANY
ASIAN NICs

1981   1982   1983   1984   1985   1986   1987

(estimate)

Sources: OECD Main Economic Indicators, IMF World Outlook, October 1987,
and Department of Finance estimates

Figure 7.2

U.S. FEDERAL BUDGET DEFICITS IN THE 1980s,
UNIFIED BUDGET BASIS, FISCAL YEARS

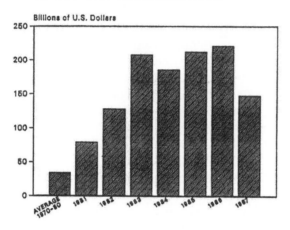

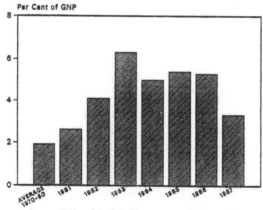

Source: Budget of the United States Government, Fiscal
Year 1988 and Data Resources Incorporated

that focus on specific developments such as import penetration and competitiveness of American industries may be helpful in identifying specific channels by which the excessive absorption is translated into a trade deficit but should not obscure the basic fact that the external deficit reflects a massive spending binge in the United States.

An alternative perspective arises when it is noted that income not absorbed by consumption must either go to taxes (T) or saving (S); substituting into (1) yields:

$$CAD = (G-T) - (S-I) = \text{Net Capital Inflow} \qquad (2)$$

That is, the current account deficit (CAD) equals government dissaving less net private saving.[4] Several points arise. First, the current account deficit is identically equal to the net capital inflow; simplified discussions that have the causality flowing unilaterally in either direction are to be avoided. Second, the implications of any given current account deficit depend upon its counterpart in the government and private dissavings terms. For example, a current account deficit caused by a surge in domestic investment has different (and less harmful) consequences than one caused by a curtailment in private saving.

Figure 7.3 illustrates the evolution of the U.S. current account in terms of the net government and net private savings rates. The increase in the federal budget deficit in the early 1980s at first was offset by the collapse in investment in the recession but later was matched instead by the surge in the current account deficit as investment recovered and private savings rates fell still further from their already low levels. Essentially, the massive inflows of foreign capital reflected in the current account deficit are necessary in order to match the excess of government dissaving over net private saving.

Finally, it should be noted that the fiscal and external deficits identified above also have had repercussions in international and domestic financial markets. Of central importance to the questions addressed in this paper is the extreme volatility shown in exchange rates. The roller-coaster ride of the U.S. dollar is illustrated in Figure 7.4; of particular interest for our later discussion is the fact that the sharp depreciation of the dollar over the past three years has occurred despite the persistence of large government deficits, and that the trade account remains in substantial deficit despite the real depreciation.

The external deficits noted above justifiably have captured the attention of many who fear their implications for the development of the world economy. In his widely read and cited <u>Atlantic Monthly</u> article titled "The Morning After," former Secretary of Commerce

Figure 7.3

COMPONENTS OF THE U.S. SAVINGS-INVESTMENT BALANCE

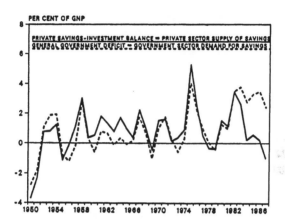

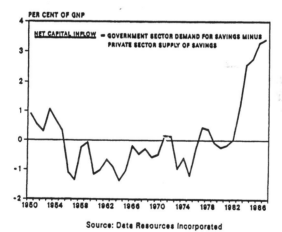

Source: Data Resources Incorporated

Figure 7.4

EXCHANGE RATE INDICES

A. U.S. EXCHANGE RATE INDICES

B. BILATERAL EXCHANGE RATE INDICES

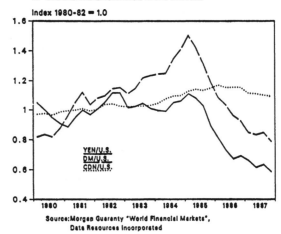

Source:Morgan Guaranty "World Financial Markets",
Data Resources Incorporated

Peter G. Peterson focuses on the "hangover" for the American economy. Others, such as Richard Lipsey in his Presidential address to the Atlantic Economic Association, focus on the protectionist pressures that use the current account deficit to promote their cause. Rudiger Dornbusch (1985, 1987) focuses on the problems of the third world debtor nations. We return to these issues after a brief review of some theoretical issues relating to the above developments.

3. THEORETICAL MODELS OF INTERNATIONAL ASPECTS OF FISCAL POLICY

The discussion of fiscal policy in the preceding historical overview focuses on the government deficit. Conventional macroeconomic theory recognizes two quite separate effects of public deficits and debt. The first underlies the now-traditional fiscal stabilization role of attenuating cyclical fluctuations in the economy; the second is the potential deleterious longer-run effects on income and welfare. Not surprisingly, much of the debate over the current deficit arises because of differences in viewpoints about the relative strengths of these two effects.

In recent years there has been a resurgence of interest in the Ricardian equivalence proposition--resurrected by Bailey (1962) and Barro (1974)--that the deficit is irrelevant, since bond-financed and tax-financed government expenditures are equivalent. Essentially, the method of financing government expenditure simply amounts to changing the time-pattern of government claims on the private sector without changing the present value of those claims. Thus in the neo-Ricardian debt-neutral world, deficits neither play a role in short-run stabilization nor give rise to a burden of the debt. Government spending matters, but the method of financing it does not.[5]

In the absence of Ricardian equivalence, government debt is treated as net wealth by the private sector and hence government deficits stimulate private consumption at the expense of private saving. Therefore, in this case the conventional stabilization role for deficits arises, and there is also a "primary burden of the debt" that arises because government debt replaces claims on productive capital in portfolios of domestic agents; there are long-run costs associated with a high debt-to-GNP ratio in the form of reduced potential national income.

In what follows I adopt the consensus, non-Ricardian view that deficits and the debt matter, but it is useful to emphasise at the outset that "the deficit" is only a summary statistic--it hides as much as it reveals, and its value as a summary statistic is directly related to the care with which it is used. For example, economists (and one hopes, policy makers!) have learned that to avoid "pro-cyclical fiscal policies," focus should be on the structural deficit rather than on the measured deficit. Similarly, concerns about the long-run effects of deficits do not mean that one should be indifferent between the relative contributions of tax increases and expenditure cuts in any deficit reduction plan. On this ground the neo-Ricardian view has the distinct advantage of directing attention to the role of expenditures.

3.1. The Mundell-Fleming (M-F) Model

The M-F model focused on the implications of international capital mobility for macroeconomic analysis in an open economy. The basic framework is summarized by the open economy national income identity, equation (1), which can be rewritten by aggregating consumption plus investment expenditure into private expenditure E:

$$Y = E + G + NX. \tag{1'}$$

With zero capital mobility and flexible exchange rates, net exports (NX) are zero as a condition of equilibrium in the foreign-exchange market. As a result, equation (1') reduces to the standard closed-economy goods-market equilibrium condition, with national income (Y) determined as the sum of private (E) and government (G) expenditures. Unless the exchange rate enters one of the behavioral relations--as in the famous Laursen-Metzler-Harberger effect--an open economy under flexible exchange rates can, for all intents and purposes, be treated as a closed economy. This "insulation property" can be argued to lie behind some of the 1950s arguments in favor of flexible exchange rates.

With international capital mobility, equilibrium in the foreign-exchange market does not require NX to be zero; imbalances in the trade account can be offset by capital account imbalances. When international capital flows are sensitive to the domestic interest rate, manipulation of that interest rate via monetary and fiscal policy creates a "degree of freedom" for the trade account to move in a manner dictated by domestic stabilization objectives. However, that "degree of freedom" comes at the expense of linking domestic interest rates to foreign and losing the insulation properties of the flexible exchange rate; the open economy becomes fundamentally different from a closed economy.

A more complete model is now required. The M-F model includes conditions for equilibrium in the domestic goods and money markets and the balance of payments. Of particular interest is the money market equilibrium condition that equates the real money supply defined in terms of home goods to the demand for real balances;

$$M/P = L(Y,i) \tag{3}$$

Since, from equation (3), fiscal policy can alter income only to the extent that it can alter the domestic interest rate, it follows that as the interest sensitivity of international capital flows increases, the effectiveness of fiscal policy is diminished. "Crowding out" is accomplished through a real appreciation, which causes a reduction in NX that offsets the rise in G.

With infinitely interest-elastic capital flows and static expectations, the domestic interest rate is determined by the foreign interest rate and the model operates much like the strict quantity theory in that equation (3) prescribes a relationship between Y, P, and M with no role for fiscal policy in influencing Y.[6] Of course, were this "fiscal ineffectiveness" result robust, it would be good news for those concerned about deficits--deficits could be eliminated instantaneously without concern for the short-term consequences for output and employment.

Superficially at least, the M-F model does not do a bad job of explaining the "facts" of the first part of the decade. The massive U.S. fiscal stimulus clearly was accompanied by some export crowding out (although national income also rose), by substantial real appreciation, and arguably by a rise in world real interest rates.

The M-F model is agnostic about how fiscal policy is transmitted to trading partners. Essentially there are two offsetting effects--the rise in American demand for imports and the rise in the world real interest rate[7]--and the transnational effects depend upon the structure of the foreign economy. Evidence suggests that the real wage rigidity in Europe meant that the real depreciation experienced in those economies had a stagflationary effect; the rise in import prices drove up European wages, thus creating upward pressure on inflation and contractionary pressure on European employment and output. Thus, U.S. fiscal policy was also a beggar-thy-neighbor policy; in that light the fiscal contraction undertaken in Europe seems all the more unwise. For many debt-ridden LDCs the deleterious effect of the rise in real interest rates swamped the rise in export earnings.

Both the European and the LDC experiences highlight the problems of the U.S. policy mix; had U.S. demand been supported more by monetary than by fiscal policy, offshore export markets could have been maintained without the accompanying high real interest rates.

3.2. Extensions to the M-F Model

The Mundell-Fleming model has been the focus of a number of critical attacks, which have provided the basis for subsequent developments in the literature. These arise from the ultra-Keynesian nature of the M-F model--specifically, its focus on short-run demand issues, its static expectations assumptions, and its absence of supply-side structure--as well as from concerns about the underlying assumptions about asset markets and the role of wealth.[8]

The so-called monetarist approach to flexible exchange rates adds a classical supply-side (i.e., with wage-price flexibility ensuring that full employment is maintained) to the perfect capital mobility version of the model. The monetarist model is therefore not interesting for the question of stabilization policy; interest instead is focused on the implications for exchange rate determination. Much of the early literature focused on the role of monetary policy and stressed the result that "the exchange rate is a monetary phenomenon"--even though fiscal policy would exert an influence on both the real and nominal exchange rate.

In an influential paper, Dornbusch (1976) combined the Mundell-Fleming analysis with the monetary approach by adding a simple Phillips curve to the M-F model. He also included a role for exchange rate expectations in the interest parity condition required by the perfect substitutes assumption; this allowed domestic interest rates to diverge from foreign rates in the short run.

In the Dornbusch model, monetary policy exerts a positive effect on output in the short run but is neutral in the long run; the only long-run effect of monetary policy is a proportionate change in the price of domestic output and the exchange rate. Further, with forward-looking expectations, the exchange rate will on impact "overshoot" its long-run response to a monetary shock. The M-F monetary equilibrium condition--equation (3) above--precludes any effects for fiscal policy on output, although fiscal policy does exert immediate and identical effects on the nominal and real exchange rates.

Michael Devereux and I have expanded the Dornbusch model to allow a richer potential role for fiscal policy (and other real shocks). We admit an influence of the exchange rate on the deflator in the definition of real balances and also of the terms of trade on aggregate supply. In our model output is affected by fiscal policy both in the short run and the long run; the latter is due to the rise in the relative price of the home good and hence the fall in the producer real wage. Further, in this model both fiscal and monetary policy must cause either nominal exchange rate overshooting or perverse domestic interest rate effects.

These results suggest that real shocks can be a potential explanation of high exchange rate variability. Further, both the potential for output effects and the prediction that the initial real appreciation will be followed by some real depreciation (as the overshooting unwinds) conform with the historical experience described above; nevertheless, from this perspective the extent of the fall of the external value of the U.S. dollar over the past three years remains a puzzle.[9]

3.3. Asset Market Effects

The Dornbusch model and the extensions discussed above all follow the M-F tradition of describing asset equilibrium solely in terms of equilibrium in the money market (implicitly assuming all non-monetary assets are perfect substitutes), and of omitting wealth as an explanatory variable for money demand. Thus, the models abstract from portfolio balance considerations and from the wealth dynamics arising from trade account imbalances.

Of course, implicit in the treatment of wealth in the M-F model is the conventional, non-Ricardian view that government deficits can alter total national saving and hence influence the external deficit, but the underpinnings of private sector saving behavior that support such a conclusion are not well specified. A series of recent papers by Jacob Frenkel and Assaf Razin addresses this issue and provides an extensive discussion of how market restrictions arising from overlapping generations models give rise to non-Ricardian behavior with optimizing agents. These papers are brought together in Frenkel and Razin (1987.)

Penati (1983) surveys the conventional portfolio balance literture. As he shows, if domestic and foreign currency denominated bonds are sufficiently poor substitutes, the monetary result that fiscal expansion leads to an appreciation can be reversed; this effect can be exacerbated if the country is a net debtor, since in this case a depreciation can increase real foreign liabilities and thus, further reduce the demand for domestic output. These considerations typically are thought of as being more relevant to the situation of key LDC debtors but may become more relevant to the United States as its net asset position deteriorates; "saturation" of foreign portfolios with U.S. dollar assets plays some role in explaining the fall in the U.S. dollar over the past three years.

The abstraction from wealth dynamics in the M-F model also is not innocuous, since the short-run effects of the M-F-type analyses likely will be reversed in the long run. A fiscal expansion that stimulates the economy in the short run will also reduce the net foreign asset position of the economy over time; for long-run current account balance, the resulting deterioration in the debt service account must be matched by an improvement in the trade account, which in turn may require a fall in domestic income. Of course, in the context of full employment, the improvement in the trade account would require either a reduction in the government deficit or a real depreciation. Thus, the currently popular question of How much further must the US dollar depreciate? is ill-formed; the answer depends on how much the government deficit will be cut.

Expectations of future policies also complicate the analysis of open economy fiscal policy. For example, in the Devereux-Purvis model, anticipated future fiscal policies necessarily lead to a perverse response of current output. This result also looks like good news to those concerned about the deficit: If the government could make a "credible" announcement that the deficit was going to be cut in the future, the capital markets would cause the real depreciation to be brought forward in time so that "crowding in" would occur immediately with consequent beneficial effects on current output and employment. The catch, of course, is that after the beneficial effects have occurred, the government may be disinclined actually to pursue the deficit reduction; if that is the case, the public may not believe the announcement in the first place and no beneficial effects would arise.[10]

Fiscal policies may also induce expectations about future monetary policies that in turn set in motion reactions that undo or reverse the direct effects of the fiscal changes. Expansionary fiscal policies may, for example, create expectations of "monetization of the debt" and hence of inflation; as a result, there will be downward pressure on the exchange rate that offsets the appreciation caused by the fiscal stimulus. Alain Ize (1987) examines a particular form of this phenomenon wherein the expectation is that loose fiscal policy will lead to a devaluation at some future date as the authorities try to tax holders of domestic bonds. Capital markets will cause these expectations of future depreciation to be brought forward in time so that the initial fiscal stimulus is followed immediately by a devaluation. Ize argues that this fits the experience of "debt-ridden" economies such as Mexico in 1982, and it may also have some applicability to the recent U.S. situation. For the first time in history, a major debtor has debts denominated in its own currency, and a great deal of wealth can be appropriated from foreign (especially Japanese and German) holders of U.S. bonds via a devaluation; anticipation of that, reinforced by the slow progress in U.S. deficit reduction, could provide some explanation of the recent devaluation of the U.S. dollar.

4. RE-ENTRY: HARD OR SOFT LANDING?

The conventional analyses and the extensions discussed above all confirm what has become the conventional wisdom: current imbalances are "unsustainable"; either substantial policy changes get introduced or a dramatic market adjustment will occur to unwind them. There also is substantial agreement about what policy changes are required: a change in the U.S. policy mix, with tighter fiscal policy and looser monetary policy, accompanied by fiscal expansion in major American trading partners. This agreement has been present for some time, and most observers also agree that the longer the change is delayed the more urgent it becomes. This view is reinforced by the failure of the U.S. trade deficit to respond significantly to the almost 50 percent devaluation of the U.S. dollar in the past three years.

The argument that the current situation is unsustainable is reinforced by the fact that in 1987 a large fraction of the capital inflows into the United States were financed by foreign central banks acting to support the U.S. dollar; essentially the private foreign demand for additional U.S. dollar assets has dried up. Many observers felt that the central bank behavior represented a holding period designed to maintain stability in financial markets during the U.S. election. These same observers predicted that the honeymoon would be over shortly after the new president took office in late January; in fact, it appeared to end the morning after the election as President-elect Bush was greeted with a sharp depreciation of the U.S. dollar.

The unwinding of the imbalances can take place in one of three ways; each involves the necessary ingredient of raising the U.S. net national saving rate, the counterpart of correcting the U.S. current account deficit.

Each of the first two possibilities gives rise to a so called soft landing, which can be defined as an unwinding without significant disruption to financial markets and without a recession. One possibility is that private savings in the United States will rise. Such a rise might be induced by pro-savings changes in tax policy as a number of economists have recommended (see, for example, the op-ed piece by former Chairman of the Council of Economic Advisors Martin Feldstein in The Wall Street Journal, Nov. 21, 1988). A rise in saving also might occur as American citizens become sensitized to the decline in their future economic prospects that the mushrooming public debt portends and consequently increase their savings. (The latter involves households acting Ricardian at the margin.) While both sources of increase in private savings carry some prospect for medium-term correction, they likely will be too gradual to play a significant role in the rapid unwinding that many observers now feel is not only desirable but inevitable.

A second possibility for a soft-landing requires a change in the policy mix in the world and, particularly, in the United States. A fiscal correction in the United States accompanied by accommodation at the Fed and some overseas fiscal stimulus remains desirable. The fiscal correction is necessary to correct the U.S. dissavings problem, while the accommodation is necessary to maintain world demand to avoid a recession. This, of course, has been the theme of several recent World Economic Summits. While some favorable developments have occurred in Japan (which has begun to act like a mature creditor with absorption rising relative to income), the summits nevertheless have been notable for their failures. Essentially the United States has failed to offer credible budget proposals and so has failed to extract commitments for expansion abroad.

The possibility arises that the new American adminstration, not being a prisoner to history, will be able to offer credible budget proposals. To most observers this probably means tax increases--perhaps they have imagined that they saw movement in George Bush's lips! The Economist magazine, for example, has identified a number of tax measures that would go some ways toward reducing public sector dissaving--see Figure 7.5. However, I would suggest that expenditure cuts are not only desirable but feasible. Better control of defense spending could, for example, produce substantial savings without compromising national security. And of interest to this audience, agricultural subsidies provide room for substantial cuts. Worldwide such subsidies represent a transfer of over $200 billion annually to relatively rich farmers in advanced countries, and hence represent a massive "reverse Marshall" plan. Eliminating those subsidies would not only contribute to American national savings but would give a direct boost to a number of debt-ridden, agricultural-based LDCs.

Dornbusch emphasises that American fiscal correction is more likely if inflation surges, thus reinforcing the perceived need for fiscal contraction. More importantly, U.S. fiscal correction may be more likely now because the market--rising interest rates, falling stock prices,

Figure 7.5

HOW TO CUT THE U.S. BUDGET DEFICIT

Year-by-year reduction in the deficit ($ billions)

	1989	1990	1991	1992	1993	Total over 5 years
Raise marginal tax rates to 16% and 30%	18.7	28.4	30.9	33.4	35.9	147.4
Raise top marginal rate to 30%	9.4	14.5	16.0	17.5	18.9	76.2
Increase motor-fuel taxes by 25 cents a gallon	22.7	23.5	23.8	24.2	24.4	118.6
Impose oil-import fee of $5 a barrel	7.9	8.2	8.3	8.2	8.2	40.8
Impose broad-based tax on domestic energy use (5% by value)	16.3	17.8	18.9	20.3	21.6	94.8
Impose a 5% value-added tax, on a broad base	0	78.8	116.6	127.0	138.6	459.0

Source: Economist, September 24, 1988.

pressure on the dollar--has signalled that the "hard-landing" alternative cannot be avoided much longer.

The third alternative then is the "hard landing" that markets likely will deliver in the absence of the policy initiatives, identified above, required to deliver the "soft landing." This scenario is originally identified with Marris (1985); the ability of "muddling through" to postpone the day of reckoning meant that his message was not widely acknowledged for some considerable period. For example, writing more of the domestic situation, Tobin (1985, p. 16) argued that the rising debt was a problem

". . . [n]ot because it portends an apocalyptic day of reckoning, a calamitous economic and financial collapse neither controllable nor reversible. Too many Cassandras cried warnings about deficits two years ago, and now they have lost their audience."

The hard landing would be imposed by the market in the following manner. At prevailing interest and exchange rates, foreigners become unwilling to provide capital flows to the United States, and indeed may no longer be willing to hold all their accumulated U.S. dollar assets. These reactions would be in part due to "saturation" of foreign portfolios and in part to increased perceived riskiness of those assets becuase of expectations of American inflation and devaluation, of rising interest rates and falling asset prices, and of higher tax rates in the United States. Those expectations thus stimulate capital flight from the United States which in turn brings about the events described above; the external value of the U.S. dollar falls and interest rates rise until foreigners are once again willing to hold existing stocks of U.S. dollar assets and the trade account has adjusted. The combined effect of reduced asset values, higher interest rates, and reduced absorption by the U.S. economy plunges the world into a deep recession; U.S. net dissaving is reduced not by reduced deficits or by increased saving but by decreased investment!

From the world economy's viewpoint, a key difference between the soft and hard landings is that the former is trade expanding while the latter is trade contracting. For the LDCs, the hard-landing combination of increased world interest rates and export contraction would be disastrous. The impact of induced debt repudiation on the already fragile U.S. financial system could be a cause for alarm. The case for the soft landing policy mix is reinforced by the view that lower real interest rates are essential in the face of increased financial vulnerability of LDCs and America alike.One irony of the hard landing scenario is that at some point the unwillingness of world capital markets to finance the U.S. government deficit will cause the needed fiscal correction to occur.[11] The key question for the world economy is whether President Bush will opt for the soft landing by introducing policies that preempt the market or whether he will opt for letting the hard landing eventually dictate the policy reaction. In the latter case, the policy will have the unfortunate characteristic of being pro-cyclical for the American and world economies.

5. PROSPECTS FOR CONTROLLING EXCHANGE RATE MOVEMENTS

The main thrust of the Mundell-Fleming analysis was a consideration of the implications of increased international mobility of financial capital for traditional macroeconomic analysis. Today, almost twenty-five years after the development of the M-F model, that issue remains paramount. The extreme volatility of exchange rates since the abandonment of the Bretton Woods system provides persuasive evidence of the mobility of capital--large pools of liquid assets are switched in response to anticipations of exchange rate movements, and thus appear to amplify such movements.

It is fair to say that this volatility was not anticipated among those economists favoring a return to flexible exchange rates in the 1960s and 1970s. Often preference for flexible exchange rates was based on an analogy to daylight saving time--when an imbalance between internal costs and external prices arises, it is simpler to change the conversion factor (the exchange rate) than to submit the economy to a prolonged wage-price adjustment process. However, even if that analogy is relevant to a once-for-all exchange rate change in the context of more-or-less fixed exchange rates, it breaks down in the context of a system of market-determined exchange rates in which such rates take on the characteristics of an asset price responding quickly and substantially to current disturbances. In the presence of capital mobility, movements in the exchange rate can become the source of the need for adjustment rather than an expedient mechanism for facilitating such adjustment.

The volatility of exchange rates, the extent and persistence of apparent exchange-rate misalignments, and the havoc these exchange rate developments wreak on national economies has focused attention on three classes of policy initiatives.

The first is increased coordination of economic policies. While the above discussion provides many examples of problems caused by a lack of coordination, it is hard to be hopeful about this possibility. There are few incentives or mechanisms to impose sanctions in order to enforce any "cooperative" arrangements agreed to. Coordination poses all the risks inherent in the discredited "fine tuning" of domestic economic policy; perhaps fortunately, these same problems likely make agreement on coordinated actions unlikely in most circumstances. Cooperation is important and to be encouraged; it can prove especially helpful in providing a framework for action in "emergencies" and in responding to major external or policy-induced shocks. However, overall one is forced to the view espoused by Fischer (1987) that the major benefits to the world economy rise from improved domestic policy performance.

The second proposal involves renewed attempts to impose exchange rate targets. Proposals vary widely in detail, but they all encounter the problems discussed elsewhere at this conference in identifying equilibrium exchange rates. Further, many propose essentially to fix real exchange rates and thus fix relative prices. The wisdom of such a policy in the face of a turbulent world economy that generates substantial and sustained real disturbance escapes me, and accordingly I remain sceptical of such proposals.

The third is that such distinguished economists as James Tobin and Rudiger Dornbusch have proposed throwing "sand-in-the-gears" by imposing taxes on international capital movements. My instinctive reaction is that the problems in implementing and regulating such a system would far outweigh any benefits, but nevertheless such proposals are gaining in acceptance if only because they fill the vacuum left by the absence of alternatives.

The economics profession has underestimated the potential for integration of world capital markets to lead to disintegration of the world financial system. Globalization has created a new set of interdependencies among trading nations, making the benefits from trade more widely available but perhaps leaving the system as a whole more vulnerable. Policies are interdependent, and appropriate domestic policies are changed. The country that pursues domestic budgetary policy or such details of its tax system as the tax treatment of capital income without regard for international capital flows does so at its own peril. Of course, if that economy is that of the United States, then it also does so at the world economy's peril.

NOTES

1. I would like to thank my discussant, Steve Kyle, and other participants for helpful comments.

2. Note that business cycles were correlated internationally not because flexible exchange rates failed to provide insulation in the manner usually ascribed to them but because <u>convergent monetary policies</u> were adopted internationally.

3. These developments are discussed in detail in Lynch (1988) and Courchene (1988); in what follows I draw on their discussions extensively.

4. In going from equation (1) to equation (2), I have moved the focus from the trade account (NX) to the current account; the latter incorporates repatriation of earnings on capital. Hence, the income term in equation (1) is GDP, while the income term underlying (2) is GNP.

5. For example, Bruce and Purvis (1984a) show that under Ricardian equivalence and perfect international capital mobility, a change in government expenditure will lead to an exactly offsetting change in the current account independent of how it is financed, but a change in taxes will have no effect on the current account.

6. Note the importance for the fiscal-policy-ineffectiveness result of the specification that real balances are defined by deflating nominal balances by only the domestic price, P; if M were deflated by a price index involving e, then the appreciation following a fiscal expansion would "create room" in equation (3) for some increase in output.

7. Devereux (1987) stresses that terms of trade effects could cause real interest rate movements and hence investment to exhibit negative correlation internationally.

8. This section draws on Purvis (1985); further references are given there.

9. This section draws on Purvis (1985); further references are given there.

10. This is an example of a "time inconsistent" policy; of course, if the government could "pre-commit" itself, then the first best policy could still be followed.

11. World capital markets posed external discipline on the Mitterand government shortly after its election early in this decade, and caused that government to reverse its policy direction; in that case, however, the international spillovers were minimal relative to the current U.S. situation.

REFERENCES

Bailey, M. (1962). National Income and the Price Level, New York: McGraw Hill.

Barro, R. (1974). "Are Government Bonds Net Wealth?" Journal of Political Economy 82 (November): 1095-1117.

Blanchard, O. (1985). "Debt, Deficits, and Finite Horizons." Journal of Political Economy, 93 (April): 223-48.

Bruce, N. and D. Purvis (1984a). "The Specification and Influence of Goods and Factor Markets in Open Economy Macroeconomic Models," chapter 16 in R. W. Jones and P. Kenen (eds.), Handbook in International Economics, Amsterdam: North Holland.

_____ (1984b). "Consequences of Government Budget Deficits." Royal Commission on the Economic Union and Development Prospects for Canada, Toronto: The University of Toronto Press.

Courchene, Thomas J. (1988). "Savings and Global Imbalances", Toronto: C. D. Howe Research Institute, forthcoming.

Devereux, M. (1987). "Fiscal Policy, the Terms of Trade, and Real Interest Rates." Journal of International Economics (22): 219-37.

Devereux, M. and D. Purvis (1984). "Fiscal Policy and the Real Exchange Rate," paper presented to the European Meetings of the Econometric Society, Madrid.

Dornbusch, R. (1976). "Expectations and Exchange Rate Dynamics." Journal of Political Economy 81 (December): 1161-76.

_____ (1985). "Policy and Performance Links between LDC Debtors and Industrial Nations." Brookings Papers on Economic Activity, 305-56.

_____ (1987). "Our LDC Debt," in Martin Feldstein (ed.), The United States in the World Economy, Chicago: University of Chicago Press.

Fischer, Stanley (1987). "International Macroeconomic Policy Coordination." NBER Working Paper Number 2224, Cambridge, MA: National Bureau of Economic Research.

Fleming, J. M. (1962). "Domestic Financial Policies Under Fixed and Floating Exchange Rates." IMF Staff Papers (9): 369-79.

Frenkel, Jacob and Assaf Razin (1987). Fiscal Policies and the World Economy, Cambridge, MA: M.I.T. Press.

Ize, Alain, (1987). "Fiscal Dominance, Debt, and Exchange Rates," unpublished ms., Washington, D.C.: International Monetary Fund.

Lipsey, Richard G. (1988). "Global Imbalances and American Trade Policy." Atlantic Economic Journal (16): 1-11.

Marris, Stephen (1985). Deficits and the Dollar: The World Economy at Risk, Washington, D.C.: Institute for International Economics.

Mundell, R. A. (1968). International Economics, New York: Macmillan, London; Collier-Macmillan.

Penati, Alessandro (1983). "Expansionary Fiscal Policy and the Exchange Rate." International Monetary Fund Staff Papers (30): 542-68.

Peterson, Peter G. (1987). "The Morning After." Atlantic Monthly (October): 43-69.

Purvis, D. (1985). "International Capital Mobility, Public Sector Deficits, and the Domestic Economy." Canadian Journal of Economics 18 (November): 723-42.

Tobin, James (1985). "The Fiscal Revolution: Disturbing Prospects." Challenge (January/February): 12-16.

Comments by Steven Kyle

"Fiscal Policy, Exchange Rates and World Debt Problems"
by Douglas D. Purvis

Purvis' paper is an excellent survey of theoretical treatments of the interrelationships between fiscal policy, debt, and exchange rates. This comment will focus first on some observations regarding the time periods (short- vs. long-run) implicit in the various models, followed by some comments on the implications of the Third World debt crisis, and will then discuss the implications of some of the scenarios presented for the agricultural sector.

The paper rightly points to the excess of government dissaving as a major determinant of accumulating U.S. debt and ballooning current account deficits; indeed, this much is self-evident from national income identities, as well as being the focus of various theoretical treatments. In a static sense it is inescapable that an excess of public dissaving will swamp what is an historically low rate of private savings. However, there is a longer-run aspect to this problem that bears emphasis. It is important to bear in mind not only how we got into a situation where this imbalance occurs, but also to look at the composition of our expenditures to determine the extent to which they will promote growth and an eventual (happy) resolution of the problem.

We borrow not just because the private sector and the government are on a spending binge but because in a long-run sense, we are poorer. That is, a spending binge is only a problem if national income is insufficient to support it. We can trace the beginnings of our problem in this sense to the slowdown in total factor productivity growth dating from the late 1960s and early 1970s.[1] One underlying factor is the secular decline in investment in the U.S. economy. Not only is investment in plant and equipment lower (in relative terms) in the 1980s than in the 1970s, and lower still than in the 1960s or 1950s, but the government budget itself has become increasingly directed toward non-productive ends. Three quarters of federal expenditures now go to interest payments, transfer programs, and defense, none of which promote productive capacity in the long run. Not only that, but there is reason to believe that government transfer programs such as Social Security depress national savings by serving as a savings substitute in personal portfolios.[2] Given the slowdown in investment and growth, our consumption and savings patterns are inconsistent with our ability to pay. Models and policy recommendations that focus exclusively on static accounting identities or on shorter-term variables ignore these longer-term factors.

Another consideration that must be emphasized is the role of the Third World debt crisis. For the past six years we have required large debtors to pay back debt at a rate that has at times exceeded 5 percent of their GDP. If they are to do this (and the wisdom of it is debatable), then they must run a trade surplus. For debtors to run a surplus there is an inescapable need for someone else to run a deficit. So far that someone has been the United States, and this fact has generated widespread concern over the eventual consequences. Though we might wish for a smaller U.S. deficit together with smaller surpluses on the part of Germany and Japan, to attempt to force our own trade balance to zero is to invite a weakening of our own financial system in the wake of defaults on sovereign debt. Though the banking system is much less vulnerable than it was in 1982 when the debt crisis first became a serious problem, the loss of such a massive percentage of the capital base of our banks would put the system in a far more fragile situation where any additional shocks could cause severe problems or collapse. This implies that the prescriptions of policy models that ignore the interdependence of macro policies must be questioned.

This last point is true not only for policy toward debtor nations but also for macro policy more generally. In the 1980s and beyond the United States cannot afford to behave as if it can choose its own inflation and interest rates (to the extent that these are in fact under government control) independently of the rest of the world. A recent example of an attempt to do this on the part of a major western nation was the attempt by France to reflate in the early 1980s at a time when virtually all of its major trading partners were doing the opposite. This attempt failed dismally, with a lower franc and higher inflation as the main outcomes. While the United States by virtue of its size and the fact that the dollar is the world's principal reserve currency has a bit more leeway, the lesson is clear. We do not conduct our fiscal/monetary/exchange rate policies in a vacuum and cannot expect to achieve desired outcomes if we do.

Much ink has been expended in recent years in the debate over whether the United States will be able to achieve a "soft" landing. While I do not think that the sky is about to fall as some of the more extreme analysts seem to think, I am impressed by the fact that we are talking about such a possibility at all. That it is even a possibility says a lot about the fragility of the current situation. One fact that should be rather comforting, however, is that the lower the dollar gets, and it is a lot lower than it was, the more attractive U.S. assets become and the less likely a wholesale effort to sell them becomes.

Without trying to predict whether we will see the "hard" landing or the "soft" landing described in the paper, it is possible to foresee some consequences for agriculture in each case. Clearly, fiscal restraint, whether forced by the market or achieved preemptively by Congress and the President, is in the offing. It is likely that this restraint will contain some cuts, possibly large ones, in federal farm subsidy programs. It should also be noted that fiscal restraint will become difficult to achieve if it is delayed until the economy is on the down side of the business cycle.

As the current expansion is already the longest in post-war history, such a downturn is likely at some point in the near future. It will no longer be possible to spend our way out of it to the extent that was done in 1982, since we are starting from a position in which we have a large deficit. The fact that the national debt is still not excessively large by historical standards (It reached over 130 percent of GNP as a result of spending for World War II) is small comfort given the size of current deficits and projections that indicate that the debt will grow far larger before it is brought under control. In addition, a far larger proportion is now held by foreigners than was the case in previous years. The tendency of the deficit to grow in a downturn will increase already existing pressures to cut farm programs as would any international agreement on agricultural subsidies such as that currently being sought in the GATT.

However, in the event of a soft landing, the negative effects of this would be offset to a great extent by the lower value of the dollar in a context of increased world trade. In fact, the devaluation already achieved cannot fail to have beneficial consequences for our export activities, particularly in relatively homogeneous commodities such as grains. The monetary easing that would support growth in this scenario would contribute to lower interest rates, which would also prove beneficial for the agricultural sector.

In the event of a hard landing, the situation would be substantially worse. First, fiscal contraction would be forced by capital flight and rising interest rates, resulting in farm program cuts as in the soft landing above, but in this case exacerbated by the adverse affects of high interest rates and lack of credit. In a world of contracting international trade, the lower value of the dollar would not result in greater exports, and could create incentives for competitive devaluations as occurred in the 1930s. At the same time, domestic contraction would reduce internal demand as well. All of these factors could contribute to a resurgence of the land value crisis of the early 1980s, with the consequent effects on the ability of financially stressed farmers to stay in business.

Finally, the paper's review of proposals for reform of the international monetary system also has implications for agriculture. First, the notion of international policy coordination sounds good but is not likely to be honored in the breach. If a policy is not seen to be in the interests of the United States then it will not be implemented, regardless of the supposed benefits of cooperation. In fact, policy makers may have an exaggerated notion of how much their "cooperation" contributed to the decline of the dollar and lower interests rates in the aftermath of the 1985 Plaza Accord. It should be noted in this case that the dollar had already started its downward trend without any help from cooperation and that it would likely have continued even without the benefit of the applause of the Group of Seven.

The second proposal concerns a return to a system of fixed or targeted exchange rates. To the extent that this is achieved, it will probably be bad news for U.S. agriculture. This is due to the higher tolerance for inflation in the United States than in major trading partners such as Germany and Japan, which historically have had inflation rates well below ours. To fix or target nominal exchange rates in such a situation would result in an appreciation of the real exchange rate. Trade-oriented sectors such as agriculture would be adversely affected.

The third proposal is to throw "sand in the gears" of international capital movements by imposing some sort of tax on cross border flows. This proposal, originally attributed to Tobin, is intended to promote long-term investment and to curtail short-term movements of "hot" money following slight variations in interest or exchange rates. This idea would also prove detrimental to agriculture if the United States remains a net borrower, as it would result in higher interest rates. However, if rates became more stable as well (an open question), risk would be reduced, benefitting all sectors but especially interest-sensitive ones such as agriculture.

Finally, this paper has some important implications for agricultural economic research. Most obvious is that analysis must take into account not only macro variables but also the fact that ours is an increasingly open economy. As the structure of the domestic and international economies change, the appropriate model for economic analysis will change as well. For example, the current emphasis on short-run equilibrations by freely flowing cross-border capital movements would have to be modified if the proposal to impose taxes on these movements were implemented. The need to tailor models to changing realities is just as important in terms of the time period on which we focus. Models of portfolio balance and exchange rate tend to focus on periods shorter than the agricultural cycle. Integration of long- and short-run factors into tractable models of the agricultural sector and the macroeconomy is an important area to be addressed. However, it is clear that agricultural assets must be viewed as one component of a market portfolio that includes both national and international elements.

NOTES

1. For evidence on this point see, for example, the study by John Kendrick and Elliot S. Grossman (1980). Productivity in the United States, Baltimore: Johns Hopkins University Press.

2. See, for example, Feldstein (1976). "Social Security and Saving: The Extended Life Cycle Theory" American Economic Review May 1976.

REFERENCES

Feldstein, Martin (1976). "Social Security and Saving: The Extended Life Cycle Theory." <u>American Economic Review</u> (May): 77-86.

Kendrick, John and Elliot S.Grossman (1980). <u>Productivity in the United States</u>, Baltimore: Johns Hopkins University Press.

Contributors

Sebastian Edwards
Department of Economics
University of California
405 Hilgard Avenue
Los Angeles, California 90024

Barry Goodwin
Department of Agricultural
Economics
Kansas State University
Waters Hall
Manhattan, Kansas 66506

Thomas Grennes
Department of Economics and
    Business
North Carolina State University
P.O. Box 8109
Raleigh, North Carolina
27695

John Kitchen
U.S. Department of Agriculture
Economic Research Service
Room 1140 NYAVEBG
1301 New York Avenue, N.W.
Washington, D.C. 20005

Steven Kyle
Department of Agricultural
Economics
Cornell University
Warren Hall
Ithaca, New York 14863

Catherine Mann
The World Bank
1818 H Street, N.W.
Washington, D.C. 20433

L. Paul O'Mara
Australian Bureau of
Agricultural and Resource
Economics
GPO Box 1563
Canberra, A.C.T. 2601
Australia

Lawrence Officer
Department of Economics
University of Illinois
College of Business
Administration
Box 4348
Chicago, Illinois 60680

Douglas K. Pearce
Department of Economics and
    Business
North Carolina State University
P.O. Box 8109
Raleigh, North Carolina 27695

Douglas D. Purvis
Department of Economics
Queens University
Kingston, Ontario
Canada K7L 3N6

Index

Printed and bound by CPI Group (UK) Ltd, Croydon, CR0 4YY

23/10/2024

01778240-0017